I'm Afraid of That Water

# I'm Afraid of That Water

A Collaborative Ethnography of a West Virginia Water Crisis

edited by
**LUKE ERIC LASSITER
BRIAN A. HOEY
ELIZABETH CAMPBELL**

in collaboration with coauthors
Laura Harbert Allen, Jim Hatfield,
Trish Hatfield, Emily Mayes, Joshua Mills,
Cat Pleska, Gabe Schwartzman, and Jay Thomas,
with an afterword by Angie Rosser

West Virginia University Press / Morgantown

ISBN
Cloth 978-1-949199-36-9
Paper 978-1-949199-37-6
Ebook 978-1-949199-38-3

Library of Congress Control Number: 2019044545

Book and cover design by Than Saffel / WVU Press
Cover image: An environmental enforcement boat patrols in front of the
chemical spill at Freedom Industries. Color-enhanced image from an original
photo by Foo Conner. Licensing: CC BY 2.0

*For the Elk River*


*All royalties for this book are donated to*
*West Virginia Rivers Coalition, Charleston*

# Contents

## Part III
## On Making and Remaking Community

# Introduction

Elizabeth Campbell, Brian A. Hoey, and Luke Eric Lassiter

Nothing was particularly distinctive about the ninth day of January 2014 in Charleston, West Virginia, except maybe the wild fluctuations of the season's temperatures, which had been all over the place. It was 6 degrees on the fourth of January, followed by 58 degrees on the fifth, then –3 degrees on the seventh. On the ninth, temperatures were projected to settle back into the more reasonable range of low 20s to high 40s. But that morning, those who lived a few miles away from the Freedom Industries chemical storage facility on the shore of the Elk River began to notice what those close by had been living with for at least a few days: a strangely sweet smell in the air. By the afternoon, the smell had made its way into the municipal water supply. Just before six that evening, in front of a crowd of reporters and backed by representatives from state and federal agencies as well as the water utility, then-governor Earl Ray Tomblin declared a state of emergency and issued a "do not use" order for residents of nine southern West Virginia counties—including Kanawha, home of the state capital. No one was to use the water for anything other than flushing toilets. It was a chilling warning: "Do not drink it," the governor sternly intoned. "Do not cook with it. Do not wash with it. Do not take a bath in it." Industrial chemicals had leaked—and were still leaking—into the Elk River from Freedom's "tank farm," overwhelming the region's main water processing plant just downstream.

This book tells a particular set of stories about that chemical release and its aftermath. We begin with one story collected by Elizabeth (Beth) Campbell during the oral history phase of a project that would eventually unfold into this book, and that would come to involve a team of oral history researchers, writers, community activists, and academics. But more on that follows.

---

In late July 2014, I (Beth) was running late (as ever) and grabbed one of the old Zoom H2 recorders on the way out the door. I was fairly certain I had the one that worked, but I gave it a quick test when I got into the car. It did work, fortunately, so I backed out of the driveway and made the short drive downtown to the lovely historic townhouse community where Rebecca and her husband, Ted, were nearly finished restoring one of the century-old row-homes. Rebecca had agreed to be interviewed about her experience of the Elk River chemical spill in Charleston and the water crisis that followed.[1] We'd been looking forward to her interview: in addition to being a thoughtful and articulate person, she was also a young mother. At the time of the chemical spill, her first child was a toddler, and her second was on the way.

Rebecca Roth has lived in Charleston since 2009. She was born and raised in southern West Virginia, about an hour and a half from Charleston. She works part time as a grant writer and takes care of their two-year-old daughter. Their second child was due in early August of that year, 2014. We talked for a little while about shared friends and acquaintances and about the many layers of difficulty we experienced during the water crisis. I began the interview by asking Rebecca to go back to the day of the chemical spill, or the days immediately before, and describe when she first knew or suspected that something was going on.

She paused to look at her daughter, then turned back and began to speak. Although there were some reports of chemical smells and water issues in the days leading up to the announced spill, she'd had no indication that anything was amiss until January 9. That evening, she recalled, "I was in the process of putting my daughter to bed and I got a text from one of my friends. It said, 'There's a water emergency and don't drink the water.' It was a group text to lots of folks. So at that point I went on Facebook, also on my phone, and started to see some of the reports from the *Gazette*, the local TV station, and realized that something serious was going on."

After she put their daughter to bed that night, she went downstairs. Ted was just starting to hear the news, too, and they decided he should go out and get some bottled water for the family. But when he tried to leave the neighborhood, he immediately ran into a major traffic jam—the likes of which had not been seen before in this neighborhood—and never made it to the store. "It was just as well," Rebecca said, "as the stores were probably sold out of bottled water at that point." (In fact, they were.) At the time, they were in the process of renovating the townhouse and hadn't yet moved in. Late that evening, Rebecca remembered there was some expired bottled water at the townhouse and drove across town to retrieve it. No one really knew much of anything at that point, and she just hoped that the water would be enough to carry them through the next twenty-four hours or so.

But by the next morning, it was clear that the water would not be available for much longer than just a day or two. "The next day," Rebecca remembered, "I just was getting more and more uneasy not knowing what was going on, and how it was affecting our young child and the baby I'm carrying. So we left town and went to stay with my mother-in-law, who is about two hours from Charleston. That growing feeling of unease and uncertainty and wanting everything to be very safe for our kids led us to leave Charleston." The next day, they left town and did not return until we were all informed that the water was safe to use again—even though, as we later discovered, it may not have been.

I remembered that unease as well. At the time, Eric and I lived in a historic neighborhood called Edgewood, which is just about a mile northwest of the Elk River, and a few hundred feet higher. I still remember the smell that morning when I went out to get the paper. It was heavy in the air, and I stood there for a few minutes, trying to identify the smell. What was it? Tires? Rubber? Licorice? By then we'd lived in Charleston for more than eight years and had grown accustomed to the different chemical smells that occasionally wafted through the valley. Fishy smells were common, as were bleach smells, sulfur smells, and a host of undefinably odd rotten smells. But this smell was a new one. It reminded me of a tea I used to drink called Morning Thunder, or of car tires, with a kind of sweet, anise edge to it.

I bent to pick up the paper, eyeing the coverage of the governor's State of the State speech, which he'd given the night before. State workers and teachers might get a pay raise this year, I noted, wondering if that raise would extend to those of us who worked in higher education. I turned and went back inside.

I thought back to that day as Rebecca talked, remembering how uncritically I had thought about the smell that morning. When we were finally made aware of what had happened, I experienced terrific anger, a building sense of helplessness and powerlessness and rage. How would I have felt if I had had a young child? If I had been pregnant? I asked Rebecca to talk a bit more about her personal feelings at the time, whether *unease* captured the crux of the experience for her, or whether there was more than that.

She looked at her daughter again and said, "I just felt this growing urgency of needing to get out of town and not be anywhere around it. I was really nervous about what it was doing to my health and the health of my family. You know, I don't think it was until later that I really was angry. I think at the time I was just so scared about what was going on."

They stayed away from Charleston until the call to flush the water system came.

She wasn't clear on exactly how long they were away. She frowned and said, "My memory is sort of blurry about that time. But we kept watching online to see what was happening. When the governor held press conferences, we would try to watch. Then we heard that the part of town we lived in had been called to start 'flushing the system,' and there were certain instructions to follow. And so at that point we gathered up our things, got back in the car, and drove back to Charleston to do the flushing process. My husband actually was the one who did the flushing, but we followed the directions pretty closely."

"It wasn't until later," she continued, "that we found out about how much was unknown and that the flushing directions were thus incomplete. For example, some of the scientists who worked on analyzing the chemicals involved [later said] that you weren't supposed to have kids around when you did the flushing. But that wasn't in the written directions we were given, so we didn't know that. So during part of the flushing process, my husband had our daughter with him, so that was a very, again, scary feeling to think that we had inadvertently endangered her. If keeping her away while we flushed the system had been part of the directions, we certainly would have followed it."

"We would smell the odor at different times, too," she said. "It would come and go, and that added to that feeling of unease, which has stayed with us for most of these last few months. That makes it a very uncomfortable way to live. It's not just what I keep describing as 'camping conditions' that changed our day-to-day lives. Cooking and cleaning with bottled water, rigging up bottled water showers and baths, driving to laundromats twenty miles away. It's this uneasiness that you don't know what's going on. We don't know what the long-term effects are going to be. We don't know what the short-term effects are going to be. We're just uneasy, and to have to live with that is a very difficult—it's a very difficult thing."

Rebecca's words reminded me of the general unease of that time and of the weeks and months afterwards when that smell would keep coming back, continually raising concerns about whether or not the water was actually safe. I asked Rebecca if, once they returned, she developed some kind of daily routine to deal with the water situation.

"Every single decision that was related to water," she said, "became a big deal. I had never been conscious of what a big deal water is. In the United States, in a town or urban environment, you don't really think of the water. You trust it. I've had times in my life where I've been in more rural environments, and even the water awareness that you have in those locations was nothing compared to what it was like after January 9. Every single aspect of the day had to be thought through. So you wake up, and you want to brush your teeth. Well, you need to go to the store to get more bottled water. But the stores were often out of bottled water. So you had to carefully plan your trips to the store. Where were you going to get water? We had friends from out of town saying, 'I'm coming through; do you need any water delivered to you?'

"And then you would go to have breakfast. Well, are you going to cook with this water? No, we decided we weren't going to cook with the water. So instead of boiling an egg, you know, maybe you had cereal and maybe you had cereal in a plastic bowl that you would throw away instead of in a bowl that you would wash. My husband actually did dishes several times in bottled water. We were just so scared of what was in our tap water.

"To take a shower, I definitely didn't shower as often as I did before the water crisis, and we finally ordered a camp shower and hung it in the shower, and we would boil water, heat bottled water, on the stove and then carry it upstairs, put it in the camp shower and then have this trickly drop of water to shower with. To bathe our daughter we got a plastic bin from Lowe's and again boiled the water, got it to the right temperature, put it in the bin, and the bin was in the tub, and then she would splash around in the bin.

"And to do laundry? There was no information about how we were being affected by whatever exposure to water or the air, so we had no way of evaluating well this is a more dangerous activity than this, let's cut down on our risks. But we don't know which activities are the most dangerous. So to do laundry we were either going to relatives out of town you know every weekend or every other weekend, or we went to Saint Albans, which is about half an hour from Charleston and on a different water supply. And I took our laundry there and did it for weeks and weeks. But when you're pregnant and your belly sticks out and you have a two-year-old to take care of, all of this is just not something that is sustainable over the long term. So the laundry and the dishwasher, running the dishwasher, were things that for us only lasted a few months and we were back to using tap water."

Rebecca was quiet for a few minutes, then she continued: "We got our water tested from one of the companies that was doing the testing in the homes, and it said the chemical was not detectable. I kept waiting for there to be something, some signal that would let me drop this feeling of discomfort, but everything that I thought might give me that feeling didn't. So we got the water tested, and even though the results came back 'undetectable,' we had learned that detectability is only at a certain level. There's already a lot in the media about the CDC [Centers for Disease Control] level, what they said was safe, but where did that information come from? There was controversy about that, too. Then we learned that although they were testing for this one particular chemical, the actual leak had involved a mix of different chemicals, not all of which could be tested for. And no one—not the CDC, not the chemical companies, not independent scientists—knew whether the mixed chemicals might have different health effects than the individual chemicals. So I thought I'd feel better after the water was tested, but I really didn't feel better about using the water.

"But it was just not sustainable to use bottled water for everything. We started hearing from some of our friends about some pretty intense water filtration systems, and they were very, very expensive, maybe three thousand dollars or so. One of our friends got that system, and so we started going to her house to bathe our daughter, and I would take a shower. We even did laundry at her house a few times.

"But we just couldn't, really, the more pregnant I got and the more time went on, it just was harder and harder to keep up all of these tap water work-arounds. My husband started using the tap water to shower. I still had these lingering feelings, like I was being too risky with our health, but I just couldn't do it anymore. We were using tap water for the dishwasher, and we were going to our friend's house for showering and bathing, and we did that for a long time. We were also in the middle of renovating our house then, but that really got pushed back, everything got pushed back, because we were spending so much energy on just water.

"All the things that I think any middle-class family anywhere just takes for granted, like let's go out to dinner on Friday night. Suddenly we were saying, okay, well, where's that list of restaurants that are using bottled water for cook-ing? Where's the list of restaurants where they're serving bottled water for drinking? And those were the restaurants that we were going to. And we would see all our friends there because that's what most of the people we knew were doing to get through this. And as the months went on, I think people just got so tired. To constantly be asking people where the water came from. Up to that point, I mean, we had been committed to doing things like buying local and eating local food. And suddenly to have so much doubt about that, too—Well, is the chemical in the soil now? Is it still in the water? We want to give Lucy whole foods, so you get vegetables and fruit at the store and then you have to rinse them off. But you don't know if the tap water is safe, so now we're using even more bottled water. So our daily lives were completely disrupted by what was going on."

I asked her if anything else had changed at that point. It was July then, six months after the spill. They had been using the tap water for dishwashing and clothes washing, and I wondered if they were drinking or cooking with the water yet.

Their families had pitched in to give them one of the whole house filtration systems as a gift, and they had installed it in the townhouse, where they now lived. "So we have a huge tank downstairs that I can show you," she said. "It cleans all the water in the house. It's the kind of tank that people who live in superfund sites get. Or if somebody gets referred by a doctor because they have cancer and they're trying to have very, very clean water, this is what they use. So we got that, and then there's a separate filtration system that goes just to a separate faucet at the kitchen sink, and that's what we're using for drinking water.

"And even that," she continued, "even that. It took a few days, a week, before I felt like I could use the water. I still, again, I still feel some discomfort even though we're doing all these things and taking all these steps. That feeling of doubt is still there because it really seemed like people just don't know what the risks are. So I feel like we did the best we could for as long as we could and that's all we can do. And I just feel like so many people now share this common experience of feeling that way, doing the best they can for as long as they can. And after a while you just—each person's place is different—but you reach this point where you just can't do it anymore, and people, I think, reach that point at different times.

"It's the worst feeling in the world to dwell on it and think I'm endangering my family because I live here and this is the water that we have to use. So I definitely understand other people not wanting to talk about it, you know, wanting to ignore it and pretend that everything is okay now. I mean it might be, but we really don't know, and there's no way to get around that for me."

We both watched Lucy for a moment, who was eating (and playing with) a small bunch of grapes. Rebecca had described the destabilizing nature of this experience in a way that rang so true. "One of the things I think you've really done in a way that I haven't heard others do," I said, "is to articulate this *dis*-ease. It's not just *un*-ease; it's this dis-ease with which we all live now. And I really appreciate that. You mentioned friends at one point, and I imagine their experiences were similar. How has this crisis affected some of the people you know?"

Rebecca thought for a moment and then said, "I think in some ways my personal reaction to it has been more intense, but I chalk that up to being pregnant as well as having a young kid. So, I think that it's a little easier to be even cavalier about it if you're not of childbearing age, [if] you're not looking out for young kids. Or if you have the ability to live somewhere else for a long period of time and wait it out and think, well, the half-life of this chemical has to mean that I can come back at a certain point and things will be okay. You know, I'm not trying to say that in any kind of mean-spirited way. I think that everybody has to deal with it, and that's not an easy thing to do. I just think that because I am pregnant I really felt very strongly about this issue. I just wanted to protect my kids."

Another question came to me then. Rebecca and her husband are well known and well respected in Charleston, they have a host of family and friends nearby, and they both do work that enables them to make an impact on our city and state. I wanted to know if they now felt differently about this place and their lives here. I feared that the question might be too difficult, but it seemed too important to ignore. I decided to ask. "You and Ted are important members of this community," I said, "and you seem pretty well settled. I imagine before all of this you figured you'd stay here forever. How do you think about that now?"

As I'd come to expect, Rebecca thought for a minute before responding: "I really like being in Charleston. My husband grew up here, and there are so many things that we really enjoy about Charleston and the quality of life we get to experience here. Having both my mom and his mom in a two-hour driving radius is fantastic. We have a great support network of friends. We have a great network of working relationships. We've been very luck to find jobs that are both challenging and rewarding. That's a lot to give up if we went somewhere else. But we did have those conversations, and I think a lot of people did after this. You just have to weigh all of that against your health, and that's a really tough call to make. I mean, I think that's why I was so set on trying to get the water filtration system because I thought that's a long-term investment that is going to keep me here and make me feel at least a little safer than I had been feeling.

"But yeah it definitely shook us as far as making a future. We wanted to renovate this house and live in a great neighborhood, and the school that Lucy will be able to walk to is a fantastic public school, and we were very excited about it. And I always like to cheer for West Virginia and love to try and talk other people into living here. That was a real, you know . . ." She pauses. "It was kind of a crisis of faith. Because suddenly, I just didn't even want to be an advocate for West Virginia. I couldn't say to people, 'Yeah, this is the best place in the world to be, to live, and to work, and to raise your family. But don't drink the water.'"

Her words hit home for me. Because the crisis—all of what led up to it, how it was handled, and now what had (or better, hadn't) come out of it has been hard for us, too. It was my turn to think for a minute. "We're not from here," I said, "but we've been welcomed here. And there's so much we love about being here. But this has really shaken us, too. It's challenged how we feel about being here, about the long-term possibility of staying. It's made us think differently about this place." I asked her if that had also been true for her. Did it make her think differently about the place generally? Or if it had, how?

Rebecca was very quiet then. "It really is a hard thing to talk about." I said that I struggle with it, too. We shared more silence. "Maybe," she said, "we can come back to that."

I asked another question. It was along the same lines but a little less deep. Still, it was a tricky question to ask. Eric and I had lived in West Virginia for nearly nine years then, and I was acutely aware of how questions like the one I was about to ask sound to natives when they're asked by outsiders. "Has this made you think about our leadership differently?"

She laughed. "I guess I want to give people in leadership roles the benefit of the doubt because, you know, this was an extremely trying situation. But I also feel disappointed that more hasn't come out of it. I don't know if you saw what I said in the legislative hearing, but I really feel like this is a situation where leaders should say, 'We're going to pull out all the stops and do whatever it takes, and we're going to tell the rest of the world that we're going to have the best, cleanest water anywhere in the country.' What an opportunity that was for us. To me that would have been tremendous leadership, and that's really what needed to happen. But it didn't. So that was very disappointing."

We'd been through all of the questions on my list, and I told her that we were nearly finished. I did want to ask one question again, though: "Has this crisis made you think about this place differently? Do you feel differently about living here?"

She answered this time: "It really is disappointing that in the capital city of the state that I was born in and grew up in and always thought I wanted to live in forever that I've ended up feeling unsafe."

I identified with what she said. Even though I wasn't born here and didn't grow up here, I also thought I wanted to spend the rest of my days in Charleston. But the January 9, 2014, chemical spill and water crisis shook me, too, and the aftermath made making a future here difficult to imagine.

---

Rebecca's story sets the stage for the kind of story we hope to convey about the Elk River spill and its aftermath. We want to emphasize at the outset that we do not claim to tell the whole story. But we do intend to bring the reader persuasively close to the scenes and subject matter even while acknowledging that what we have produced is not *the* story but rather a thoughtfully curated *collection* of stories. In fact, as we finish this book many years after the spill, the number of accounts is burgeoning. In 2016 the U.S. Chemical Safety and Hazard Investigation Board released its final report into the "2014 Freedom Industries Mass Contamination of Charleston, West Virginia Drinking Water"; the spill, subsequent contamination, and repercussions have been chronicled by national and international media outlets; and our local paper, the *Charleston Gazette-Mail*, has exhaustively followed the story (see, e.g., the note references in chapter 1). Dissertations are being written; films have been made. Each of these accounts tells a different kind of story, and the ones we present here explore the spill and water crisis from the perspectives of some who experienced it. Recognizing that no story is ever complete, our hope is that what we present here will help readers gain a stronger sense of what this experience was like for us and how it has changed us. Our experience of that mass contamination event is grounded in the very particular context and history of this place and has pushed forward an equally particular set of community activisms and changes.

We also tell these stories in a particular way. This is, in many ways, an ethnography, a specific research method and literary genre that has among its main goals to describe the on-the-ground experience of an event or place. As writers of ethnography, we concern ourselves with layered histories in careful contextualization and offer some insights beyond the immediate particulars of the local; we suggest possible links for ourselves and others to trace to places both near and far. This particular ethnography is rooted in the oral histories of citizens who were—and in many ways, still are—on the receiving end of the water crisis, rather than on the agendas, perspectives, and experiences of governmental and company officials, which have been covered extensively in a number of other outlets. Because this book is a collaborative ethnography written by a variety of academic- and community-situated authors, it has two very specific audiences: academics who read ethnography (e.g., faculty and students interested in, say, disaster studies) and a local audience of West Virginians who experienced and who are trying to come to terms with the January 2014 chemical spill and its ensuing water crisis. In a more general sense, this ethnography is also directed at a book-reading general public who might be interested in the particulars of this event as well as how it might relate to larger issues of disaster and its aftermath.

This book has been collaboratively conceptualized, researched, and written by people across various positions in both academe and local communities: readers will encounter a range of voices in these pages as compelling and different ways of speaking to both personal and shared experiences and concerns. In a four-year period, more than fifty people, most of whom directly experienced the crisis, contributed to and helped to write this book. Participants—ethnographers, interviewers, interviewees, academics, community activists, and other contributors and collaborators—range in age from young adult to long retired. Voices of administrators, artists, chemical engineers, faculty, and farmers are included, as well as those of homemakers, lawyers, service workers, students, physicians, teachers, and more. They came from the working, middle, and upper classes. Some came from families who lived in this area before statehood; some only recently arrived.

We describe exactly how all of these people came together in chapters 1 and 3; for now, suffice it to say that, as in many collaborative projects, it has been an organic and emergent process. It all began, really, with the frustrations of a chemical engineer. A connection between that engineer, leaders of a local nonprofit organization, and an academically situated researcher came next; these connections led to a collaborative oral history project. From there, collaborations with other researchers and documentarians emerged. Two graduate seminars came next, along with a host of conversations and potluck dinners; then, finally, came a larger collaborative process that gave rise to this book.

Readers will encounter everything here from folksy aphorisms to academic theory, from personal soul-searching to comparative analysis. The overall book narrative generally moves from a description of the specific context leading up to the spill, then to how it might be understood more generally, back to the particular experience of the spill itself, and finally dealing with its aftermath. It explores the range of emotional, existential, and activist responses that rose to the spill, but it also presents a bird's-eye analysis of how this event and its aftermath share qualities with disasters elsewhere and reflects on where we are now. Although there is a general pattern in how the stories are arranged, we want to emphasize that the text itself does not progress in a traditional fashion or unfold in a smoothly narrative way. Human experience is rarely a complete, neatly packaged thing, and this book in many ways deliberately reflects this partiality. Again, the stories here present experiences of this event from a range of perspectives. Our goal is for readers to come away with a sense of how different people experienced—and were changed by—the 2014 Elk River chemical spill and its ensuing water crisis. If that is your goal as well, we recommend that you approach this book with an openness to the broad range of people and experiences that make up our many and different communities.

We should note, too, that because of the book's particular history and approach (again described in chapters 1 and 3), the authors involved in this work come from a range of different positions and backgrounds—some of us are academics, some are current and former students, others are professionals, and still others are community activists from various sectors. We also possess a range of varying skills and expertise. The Appalachian natives of our group, for example, possess an expertise about living in this region that far exceeds those of us not from here; the long-haul activists bring expertise on what is (and is not) possible here, much more than those of us more tangentially engaged in community change; and the academics bring expertise in their fields that reflects years of specialized study. As individuals, none of us possess all of these things, but as a group we can bring together our different positions and understandings such that, through a collaborative frame, we can offer a broader, and deeper, understanding of the event and its aftermath that none of us could do alone.

Readers should thus expect these different backgrounds, positions, skills, and expertise to emerge differently in different chapters. Chapters describing oral histories written by community activists were meant to be written in close-up prose emphasizing experience first and foremost and were purposefully *not* written within, for example, an anthropological framework with its attendant theories of disaster experience. Luke Eric Lassiter explains in his chapters how we brought Brian A. Hoey into the project to provide just that kind of anthropological analysis, which Hoey offers in chapter 2 and an interlude. Readers more interested in academic or theoretical analysis should, then, zero in on Hoey's discussions; those more interested in the logistics of collaborative ethnography, on Lassiter's chapters; and those more interested in descriptions of community experience, on those chapters by the likes of Trish Hatfield or Jim Hatfield, who also brings his expertise as a local chemist.

Still, the chapters are not meant to be read alone without their connection to the larger whole. They are not, for example, an edited collection of essays in the conventional sense. All of the chapters have been worked and reworked in the context of a team-written ethnography and, we should make clear, within theoretical currents of contemporary collaborative ethnography, which purposefully assemble various positions, voices, and perspectives as diversely located descriptions (more on this in chapter 3). Thus, chapters may be very different from one another, even as they work within the context of the same project. Our hope is that our collaborative approach to this book will point readers to both this disaster's uniqueness and its commonalities with other disasters like it.

To be sure, experiences of disaster in the Kanawha Valley are neither unique nor uncommon; it is possible—even likely—that readers' interest in this book will be piqued and informed by their own experiences of disaster elsewhere. We hope that this book will connect with larger conversations into growing regional, national, and global concerns about the human necessity of regular access to clean, safe water. The crisis we explore has left us with two critical understandings: first, that people's experiences of disaster always emerge within a historical context (and ours was no exception), and second, that our current experiences can galvanize collective action that both reflects that history and points to possibilities for change.

## Note

1.  Rebecca Roth, interview with Elizabeth Campbell, July 22, 2014. All quotations from interviews in this book have been set in dialogue with the main text, and are not set off as block quotations, no matter how long.

Part I

**"I'm Afraid of That Water": A West Virginia Disaster and Water Crisis**

# CHAPTER 1

# The Elk River Spill: On Water and Trust

Luke Eric Lassiter

"I'm afraid of that water. I've never had anything that's happened in this chemical valley affect me the way this has affected me."[1]

It was late September 2016, and Sue Davis, a local activist in the Kanawha Valley—the region in and around Charleston, West Virginia—was addressing the U.S. Chemical Safety and Hazard Investigation Board (CSB), which was meeting in downtown Charleston. The CSB had just released their long-awaited investigation report of a chemical spill that, more than two and a half years earlier, had contaminated our drinking water here.[2] During a public comment period, Davis delivered her thoughts in a defiant tone, making many in attendance visibly uncomfortable. Many looked down as she talked, as if embarrassed. But her words also resonated with many others in the room who seemed to share her anger; they nodded their heads in approval as she spoke. Indeed, as others would make clear that evening, the memory of the event was still fresh in many people's minds, including mine. A resident of Charleston then, I also remembered the event well.

On January 9, 2014, residents across Charleston awoke to an unusual licorice smell in the air and a similar taste in the public drinking water. Though some would later say they had noticed the smell many days prior, on the evening of January 9, residents were informed that the tap water in tens of thousands of homes, hundreds of businesses, and dozens of schools and hospitals—the water made available to as many as three hundred thousand citizens in a nine-county region—had been contaminated with MCHM (4-methylcyclohexanemethanol), a chemical used for cleaning crushed coal.[3] State officials traced the contamination's source to an aboveground storage tank owned by Freedom Industries, which eventually leaked an approximated ten thousand gallons of the chemical into the Elk River, just one and a half miles upstream from the intake of West Virginia American Water's water treatment plant in Charleston. The spill (or, perhaps more appropriately, "release") rapidly overwhelmed the plant's filtration purification system. By day's end on

January 9, West Virginia American Water had issued a "do not use" water order—a ban restricting the use of water to toilet flushing and firefighting that remained in effect for as many as nine days in parts of the plant's service area—and the governor had declared a state of emergency. By the next day, President Barack Obama had declared the nine counties a federal disaster area. An article in the April 2014 issue of the *New Yorker* described the event as "one of the most serious incidents of chemical contamination of drinking water in American history."[4] The Chemical Safety Board's investigation report echoed this point by noting how the event highlighted a broader and critical necessity to assess similar risks across the country.[5]

As both the *New Yorker* article and the CSB report make clear, however, the event was far from over after the "do not use" order lifted. Detectable traces of MCHM remained in the water supply for weeks after the spill, as state and federal health officials struggled with (and dodged) definitions of just what could be called "safe" when describing the public drinking water—in large part because of the lack of toxicological data for MCHM. For example, a full three months into the crisis, Dr. Rahul Gupta, then head of the Kanawha-Charleston Health Department, reported that he and his family still were not drinking from their home's water taps. Too much about MCHM was unknown, he said.[6] A survey conducted at the time, in April 2014, confirmed that, like Gupta, most people did not trust their water: only 36 percent of residents reported they were using their tap water for drinking.[7] For months after that, the local newspaper continued to report on residents who harbored reservations about their tap water.[8] And long after the water crisis was declared over, many residents continued to question their water's safety.[9] To be sure, at the time of the CSB report, Sue Davis wasn't the only one who still had serious concerns.

In his book *Disaster Culture*, Gregory Button argues that the media, government agencies, and others commonly single out disasters, be they "natural" or "unnatural," as isolated or unique occurrences. While such descriptions can often have the effect of bringing immediate attention to an event (such as when a disaster is described as the "worst of its kind"), and while they may underline the need to generalize lessons learned (as in the CSB report), they can also have the perhaps unintended effect of making a disaster seem outside the realm of what is normal. In actuality, argues Button, events like the Elk River spill—such as the ongoing water crisis in Flint, Michigan, that began in 2014—are increasingly common and not in any way unusual. Importantly, the lingering everyday uncertainty—in our institutions or in our government or in basic services, like having clean drinking water—that these disasters can leave us with can have lasting impacts far beyond any given disaster or incident itself.[10] The uncertainty associated with disaster is also fast becoming endemic in modern society.

Button's meaning of *uncertainty*, I should make clear, involves much more than how any given disaster may leave in its wake a sense of doubt about institutions or basic services; it includes a broad range of "discursive practices" involving complex constellations of suspicion, ambiguity, misgivings, and distrust "shaped by political, economic, bureaucratic, ideological, and cultural concerns" that get tied up in disasters as they emerge as sociohistorical, cultural, and political events. Button illustrates, for example, how corporations and public agencies often mobilize uncertainty to deflect blame or downplay a disaster's impact, such as when they use the "inconclusiveness" of scientific studies to "retard liability or protect polluters from government regulation or cast blame and responsibility on others."[11] The production of that uncertainty (in carefully orchestrated public relations campaigns, such as in the 1989 *Exxon Valdez* oil spill) often masks the underlying structural problems of so-called manmade disasters, in particular—structural problems that are often either ignored or glossed over by the media and by the governmental agencies meant to protect us from such catastrophes in the first place.[12]

Part of that glossing over is a topic we will return to often in this book (in chapter 2, for instance): namely, the on-the-ground experience of the spill's effects that would come to inform persistent uncertainties about safe water that still surface in conversations here in the Kanawha Valley in and around Charleston. Why do people continue to have concerns about the safety of their water, even years after the spill? What experiences inform this uncertainty, and how are they relevant to understanding the lasting impacts of this disaster—to Charleston and the Kanawha Valley, to the state and nation, and beyond? How can we understand such concerns as a critical part of this disaster's analysis, rather than as marginal or insignificant?

In sum, what if we take Sue Davis's comments—and the direct experience that informs her sentiment—seriously? Button points out that in many analyses of disasters small or large, "uncertainty is too often relegated to the realm of the irrational. At times it is dismissed by analysts, policy makers, and politicians as some kind of free-floating anxiety whose pursuit is fruitless and threatens to undermine rational discourse."[13] Currently, comments like Davis's are considered by many in the Kanawha Valley to be reasonable and well within the norm when conversation turns to infrastructure, particularly to public drinking water. Conversations about this uncertainty have spilled over into larger discursive practices between and among the local privately owned water company, government agencies, local and state government representatives and their constituencies, activists, and citizens' groups—all of whom are now pushing in various directions for control of water and how water should be viewed and understood (as safe or unsafe), owned and regulated ( privately or publicly), or consumed and disbursed (as a basic right or a privilege).

But I'm getting way ahead of myself here. Before we explore this problem in more depth in this and the following chapters, I need to first take a moment to sketch the history behind this event, elaborate how it unfolded into a water crisis, and explain how all of us involved in this book project came to do oral history and collaborative ethnography.

## Making Soup and Its Aftermath:
## A Brief Overview of The Elk River Spill

As I pointed out earlier, I remember the day of the spill quite well.[14] On the morning of January 9, I noticed a licorice-like smell in the air around 7:30 a.m. when, as I did each morning, I opened the back door to let out our outside cat, Uncle Henry. I remember, vividly actually, stepping out the door and into the yard to get a better whiff of what I thought at the time was one of the oddest aromas that I had ever encountered while living in Charleston the past nine years. It was quite unlike others I had inhaled in the night or morning air. My wife, Beth Campbell—who, like me, also at the time worked at Marshall University—noticed the smell, too, when she stepped outside to retrieve the morning paper. We conferred on the smell over coffee that morning, comparing it to those we'd experienced before—rotten eggs, fish, burning rubber, bleach, musty garlic. We came to the eventual conclusion that this smell was a new one.

Clearly, it's not unusual to catch a strange smell from time to time here, especially at night, when these traces seem to materialize most often. Many simply chalk it up to part of what it means to live in Charleston and the Kanawha Valley, a place folks here often call "Chemical Valley." How the Kanawha Valley became a chemical valley—eventually hosting various chemical manufacturers from Bayer to DuPont to Union Carbide to Dow—is a long and complicated story beyond the scope of this relatively brief introduction. Suffice it to say, though, that historians often trace the development of Charleston's chemical industry to the growth of salt production in and around Charleston in the late eighteenth and early nineteenth centuries. Before the American Civil War, salt production became heavily dependent on industrial-based slavery; by the end of the war (during which Lincoln famously admitted West Virginia to the Union as a new state in 1863), industrial salt production in the Kanawha Valley, absent that slave labor, declined dramatically. The First and Second World Wars, however, brought a new demand for products incorporating chemical building blocks found in abundance in the Kanawha Valley, such as chlorine found in salt, together with other feedstocks like coal, oil, and natural gas. Diverse manufacturers fixed their sights on the Kanawha Valley again, and by the mid-twentieth century, the city of Charleston and surrounding area had emerged as a major hub for some of the world's largest chemical manufacturers. Though that growth was relatively short-lived—the Charleston-based chemical industry began a slow decline in the second-half of the twentieth century (many manufacturers, for instance, relocated to the Gulf Coast where natural gas and oil-derived feedstocks were more abundant and affordable)—today it still remains a vital part of the local economy.[15] More than a few students in the graduate program I direct have been associated with the industry in one way or another. And as we will see, a chemical engineer—married to one of our graduates—inspired the research behind this book.

In any case, I was rather disconcerted by the morning's air quality, but admittedly, I had all but forgotten it by the time I left for the office. That evening, Beth and I were both a bit late in coming home from the Marshall University Graduate College campus in South Charleston. We had been absorbed in various faculty and student meetings most of the day—it was the week before the beginning of spring semester classes—and had heard nothing about the emerging crisis. We arrived home and headed straight for the kitchen without turning on a radio or television and decided to make a soup, a butternut squash soup to be precise. Some time later, we had just finished a second bowl each of what we decided must be one of the best soups we had ever made: sweeter, richer, more complex. And then the phone rang. We don't usually answer the phone during dinner, so we let the answering machine pick up. As we finished the last of our soup, we listened while an automated recording announced the "do not use" advisory, that West Virginia American Water customers should not drink, cook with, bathe, or wash with their tap water. One of our neighbors called next, curious to know if we had heard the news (dinner abruptly over, we answered that call). We turned on the television and learned of the day's extraordinary events.

Local citizens, closer to the spill site, were the first to respond that morning of the ninth, several calling 911 and reporting the strange licorice-like odor to the West Virginia Department of Environmental Protection (WVDEP). In the late morning around 11:00 a.m., WVDEP personnel arrived on the site of the aboveground storage tanks owned by Freedom Industries, where they discovered a "fountain-like flow" of what they were told was MCHM leaking from tank 396 into the Elk River. The Elk River, a tributary of the Kanawha River, is the sole source of potable water for most residents living here and in surrounding communities. It sources the Kanawha Valley Treatment Plant, privately owned by West Virginia American Water (WVAW) and regulated by the state's Public Service Commission. Subsequent investigations confirmed that WVAW had known chemicals were stored in aboveground storage tanks one and a half miles upstream from their raw water intake for more than ten years but had not inquired as to their contents, and so had not developed ways to detect their presence in the river or in their water treatment system. By early afternoon, WVAW employees had started to notice a strange odor inside the treatment plant. Based on what they were told at the time WVAW attempted to remove the water's licorice-like odor and taste but were unsuccessful, as the agent overwhelmed their treatment process by the late afternoon. Their information was incomplete, however: for twelve days, for example, Freedom did not tell WVDEP or WVAW that their mixture of crude MCHM also contained other chemicals. In communication with the governor's office, WVAW officials decided not to close the treatment plant's water intake (per the impact that would have on sanitation and fire protection) and issued a "do not use" order at 6 p.m. By late evening the governor had also issued a state of emergency and encouraged citizens to use the water only to flush toilets and put out fires.[16] "Do not drink it," he said at a press conference. "Do not cook with it. Do not wash clothes in it. Do not take a bath in it."[17] I went to the kitchen and ran the water. I put my nose near the tap: the origin of the licorice-like odor I had detected in the morning air was clearly in the water itself, and in its source, the Elk River. And for the first time, I also noticed that the water seemed to have a darker tinge than normal, albeit very slight, and a somewhat slick feel. Just an hour or so earlier, going about our routine dinner preparation and evening conversation, Beth and I hadn't noticed any of these characteristics, including the odor when we made our soup.

Others throughout the valley would report similar experiences. "You could say I was probably the later bloomer when it came to finding out about the water crisis," local physician Shelda Martin, would report. "I had worked all day. I came into the hospital early that morning and had worked all day until about 5:00 p.m. in the evening. I hadn't eaten, hadn't taken a break to go to the restroom, was walking across the hospital to get to my office, and I realized I was dying of thirst. So I stopped at the nurses' station on 3-South at [the Charleston Area Medical Center] Memorial Hospital to get a drink of water.

I got a drink of ice water from the patient's ice machine and noticed it was a yellow-brown color.

"I thought, 'Oh, it's probably just some particulate matter, right?' I drank it because I hadn't drank anything all day, and I'm like, 'man, this is the nastiest tasting stuff I've ever smelled!' I said, 'Who is your charge nurse?' I asked the nurses at the station, because I said, 'you need to call maintenance up here and get the ice machine and the water machine fixed—something's wrong.' And they said, 'oh, okay.' They didn't know about the water problem either. So . . . I walked over to my office . . . and I was working in my office. It's now like 6:30, quarter to seven, and one of the girls in student services came out in the hallway and said, 'Dr. Martin, I smell something burning in the building.' She and I walked the whole third floor and second floor looking for it. The smell was like a melted vanilla candle that somebody was burning, an electric candle. I can't describe the smell, but we knew something was wrong. We couldn't find it, and we called security. They came up to the third floor of our building and said, 'you ladies need to leave the building, there's a water crisis and we've had to shut the building down.' And so, it's what, 7:00 at night, and apparently this had been going on all day, and I had no idea!"[18]

The "do not use" order affected a nine-county region, which included homes, schools, hospitals, restaurants, and even the Charleston's State Capitol Complex (where the legislature had just begun its 2014 regular session the previous day).[19] At first, panic set in, with thousands of people clearing out bottled water supplies in convenience, department, and grocery stores throughout the affected region. Beth and I, for example, decided to search for bottled water soon after 8 p.m. We drove through several nearby towns and cities and then further and further out; it wasn't until late in the evening that we found water in a grocery store outside the affected region about sixty miles from our home—and even then, we found only soda water on the shelves. (We bought about two dozen liters, which we used for drinking, cooking, and bathing for several days.)[20] By the next morning, thousands of gallons of clean water—in water buffalos (a type of water tank) and as commercially bottled water provided to residents at no cost—began arriving at water distributions centers throughout Charleston and the surrounding counties. We had seen some of those trucks driving along Interstate 64 during our hunt for water that evening, a sight we found simultaneously comforting and eerie. That seemed to ease the alarm that many understandably felt, though the water distribution was not without tension, as some people's patience wore thin or tempers flared. For the most part, water distribution—managed by various state agencies, the West Virginia National Guard, and the Federal Emergency

Management Agency, among others—seemed, at first, to be well organized and executed (though many eventually pointed to notable exceptions in some areas of high-density residency and low average incomes [see, e.g., chapter 9]).[21] This relatively swift distribution of potable water was fortunate, given that the State of West Virginia admitted that it had no emergency plans in place for a chemical spill of this type.[22] Kanawha County "emergency officials, who are charged by law with chemical accident planning," the local newspaper reported, "didn't act to prepare for this type of incident, even though they had been warned for years about storage of toxic chemicals so close to the only water treatment intake serving hundreds of thousands of people."[23]

The spill itself was just the beginning of the crisis. Very little was known about MCHM at the time of the spill, especially its potential health risks.[24] And as the chemical now contaminated the entire water system, state officials and emergency responders struggled to deal with the emergency "on the spot," as it were. A growing number of people began to report various physical symptoms such as nausea, rashes, vomiting, abdominal pain, and diarrhea.[25] Based on physician reports and community surveys, some officials estimated that, by the water ban's end, more than one hundred thousand people had suffered some sort of physical ailment.[26] Many more are expected to have not reported their symptoms and thus were not included in these numbers, of course. Beth was among these. After returning home from our search for clean water, she started experiencing abdominal pain that, as she described it, "felt like needles in my small intestines." She assumed the symptoms were related to stress, but by the next morning, her symptoms expanded to include nausea and diarrhea.

Beyond the spill's effect on community health, its economic impact was also acutely felt in Charleston and the surrounding region. One preliminary report estimated that the spill cost businesses $61 million in the first four days after it occurred and at least $19 million each business day over the course of the entire water ban—though, as noted by the report, these conservative estimates did not include ripple effects such as costs for cleanup or emergency response. Of the estimated seventy-five thousand affected workers, among the hardest hit were "the lower-wage, service-producing sector," especially in the "restaurant and lodging industries [which were] . . . less likely to recover lost revenues."[27] Take, for example, those working in restaurants, which were ordered to cease operations immediately. To reopen, each restaurant had to submit written plans to the Kanawha-Charleston Health Department for how it would obtain and use bottled water for cleaning, hand washing, cooking, and drinking. Inspections accompanied plan reviews, so this took time. Over the course of the initial nine-day ban, various restaurant workers—servers, cooks, dishwashers—lost up to fifty hours of work.[28] Though several local restaurants participated in popular initiatives to help their employees recover lost wages (one, for instance, encouraged patrons to "turn up the tips"), many workers still struggled to rebound, even long after the spill incident itself was over.[29]

Restaurant owners, too, took a significant hit. Several reported that they did not receive insurance coverage. And costs for providing bottled water to customers—which many restaurants continued to do for months after the spill—were prohibitive.[30] A Charleston-based Italian bistro, for example, was still assuring customers as late as May 2014 that they were "cooking with bottled water," a promise that the restaurant owner estimated that had "cost $10,000 over four months in extra bottled water and ice."[31] Despite these obvious challenges, though, many restaurants did recover; one of our favorite eateries, for example, added a water surcharge for the many months (over a year) that it continued to use bottled water for cooking and drinking. Others did not recover. A café near our offices in South Charleston, for example, closed its doors during the spill and never reopened. Although our favorite Japanese restaurant did reopen, it never overcame the loss of revenue and increased expenses that resulted from the spill. Within three months, it closed down for good.

While restaurants and other service industries (like hotels) were busy trying to find ways to continue operations, state officials struggled with reckoning how to remove MCHM from the water supply and get safe-to-use water back to almost three hundred thousand people. At the time of the spill, the available safety data sheet for MCHM provided "little information that could be used to determine an exposure threshold." Eastman Chemical Company— the manufacturer that sold MCHM to Freedom Industries—provided several "proprietary toxicological studies," from which the West Virginia Bureau of Public Health and the Centers for Disease Control and Prevention determined a "short-term screening level of 1 ppm."[32] But several chemists and other independent scientists immediately called foul, questioning how the CDC had arrived at their conclusions and noting that, even with the Eastman studies in hand, too much was still unknown about MCHM and that the lack of clear MCHM data raised questions about whether such an exposure threshold was actually safe for consumption.[33] "The data needed to make that assessment," pointed out one chemist, "simply do not exist for this chemical."[34]

Given these shortcomings and disagreements, state officials proceeded with one part per million as the temporary threshold to begin the testing and subsequent flushing of the entire water system. After testing water samples at points throughout the region, four days after the spill, on January 13, West

Virginia American Water began flushing its system by zones. When water tests showed MCHM at or below one part per million in a particular zone, that zone's customers—residents, businesses, hospitals, schools—were instructed to begin flushing their water lines. (The zone in which Beth and I lived, on Charleston's West Side, was within a few miles of the water treatment plant and was among the earliest slated for flushing.) The three-part process took a little more than fifteen minutes and involved the flushing of hot- and cold-water taps, outside faucets, and appliances. This process proceeded zone by zone, taking several days, and ending for some as late as January 18, when the "do not use" order was finally lifted for all affected areas.[35]

The flushing process was not without problems, however. Some zones, for example, had to undergo additional flushing due to MCHM levels remaining at or above one part per million after the first attempt.[36] And numerous residents reported further physical symptoms during the flushing process, according to a telephone survey conducted by the Kanawha-Charleston Health Department.[37] "So we followed the procedure," recounted Saint Albans resident Carla McClure. "Phillip, my husband, was sitting in the living room in his recliner in his usual position as I did the flushing procedure, and he encouraged me from the sidelines. So they said be sure to also go outside and flush your outside taps. I did the whole thing. Well, at the point where I had the indoor water running, and I started to go outside, I realized I'm short of breath. I'm dizzy and I think I'm going to faint. So I sat down quickly on the porch steps just for a few minutes. It still didn't occur to me that this could be the water because no one had mentioned anything about inhalation at that point. It had all been about drinking. And I sat out there in the fresh air for a few minutes. It was very cold but I got able to go and finish up but came back in and was just sick—nausea hit me. And Phillip also. I mentioned it to Phillip, and he was also having symptoms, and that was when we realized this is connected to the water. . . . That night Phillip started this twenty-four-hour period of being just very, very sick. But we still thought maybe it was the flu because I wasn't sick like he was. But he was throwing up. He was having heart palpitations. He was sweating. You know, it was like flu symptoms we thought, but he didn't go to the doctor, and after twenty-four hours it started to clear up. That was the sickest I've seen him since . . . years earlier."[38]

In any case, state officials reported that they were carefully monitoring the flushing process, testing for the agreed-upon MCHM screening level as the water system came back online for thousands of people. As an added precaution—and two days after declaring the water safe to drink—they added the recommendation that pregnant women not drink the water until tests showed that crude MCHM was not detectable at any level.[39] Unfortunately, though, at the time officials were unaware of another chemical also present during the spill. Four days after the "do not use" order was lifted for all areas and twelve days after the initial spill discovery, Freedom Industries admitted that it added PPH (propylene glycol phenyl ether) to the MCHM mixture it acquired from Eastman: it made up at least 7.3 percent of the overall compound spilled into the Elk River.[40] This revelation, of course, created new waves of uncertainty about the water's safety and raised new questions about its long- and short-term toxic effects.[41]

Added to this, for days and sometimes weeks after the flushing procedure, many local residents (Beth and I included) "could still smell the objectionable licorice-like MCHM odor in their water even after concentration levels were reduced well below 1 ppm."[42] Customers understandably wondered about the permeation of MCHM (and now PPH) into the plumbing systems of homes, schools, hospitals, and businesses and how that might affect water safety in the near and distant future. At least one environmental engineer reported that the flushing may not have worked and that the chemicals "stuck inside pipes and hot-water tanks . . . could be breaking down into other toxic materials that have yet to be fully identified."[43] Officials countered, as did the CSB report two years later, that "the highly recognizable licorice odor of MCHM can be de-tected at very low concentrations . . . even in drinking water with high levels of chlorine," and that the flushing may have even exceeded safe guidelines, more than what was needed to be considered safe.[44] Given this, though, the lingering odor of MCHM, which some residents reported for months, served as a regular prompt for larger (and growing) questions about trust.[45] As further indepen-dent studies showed that MCHM continued to show up in home plumbing (and might still produce adverse health effects), and as some customers, such as schools, had to undergo additional flushing weeks after the initial flush-ing as a result, those questions only intensified.[46] With this intensification, a second wave of contamination, this time involving a flow of information, seemed to surface: Could anyone—from state officials to privately owned water and chemical companies—be trusted to tell us the whole story? *Could* they tell us the whole story even if they wanted to? Should we just go ahead and drink the water and hope for the best?[47]

These developing questions nurtured additional uncertainty as more about what led to the crisis in the first place came to light.[48] Take, for example, Freedom Industries. The public eventually would come to know that Freedom's founding executive was a twice-convicted felon, and that he and the company were tied to a labyrinth of shady and unethical business practices.[49] Regarding the spill in particular, inspectors reported that Freedom had developed no ongoing inspection program, including leak detection, and had knowingly violated air and water pollution laws.[50] Inspectors also reported that Freedom had taken no action to contain an obvious leak the company likely knew about until it was ordered to do so and that it actively countered developing safety and other information as it sought to spin public relations in the company's favor in the days during and after the spill.[51] Some of the most damaging evidence came later, as inspectors uncovered more about the condition of Freedom's aboveground storage tanks, originally built in the 1930s by previous owners. The CSB found no documentation of any internal tank inspections before or after Freedom acquired the site just nine days before the spill. Additionally, two months before the spill incident, an independent consultant noted in an informal visual review that the tanks were in questionable condition and recommended that they should undergo a complete certified inspection before further use. Intentionally or not, Freedom chose not to act on the warnings.[52]

That the spill likely could have been avoided is, unfortunately, an all-too-familiar story in West Virginia. The state has hosted some of the worst industrial accidents in the nation's history—many, if not most, preventable—from the Hawks Nest Tunnel disaster in the 1920s to the Sago Mine disaster in the early 2000s. Many people who live here can recite a wide range of both historic and contemporary cases, often related to coal in some way or other. An oft-heard story is of the 1972 Buffalo Creek flood, in which a coal waste dam owned by the Buffalo Mining Company collapsed and released more than 130 million gallons of sludge into Logan County's Buffalo Creek in the state's southwest region. The flood killed 125 people and destroyed nearly everything in its path, leaving thousands homeless. State officials, it turned out, had never approved the dam's construction in the first place; in addition, both the company and state were warned repeatedly that the dam might be unsafe, but both failed to act.[53]

Closer to the disaster at hand, many here also remember and recite previous chemical-related accidents. At least two incidents were fresh in many people's minds at the time of the Elk River spill, including a 2008 explosion at Bayer Crop Science Division in Institute, just west of Charleston (that explosion rattled the windows in our house, a little more than eight miles away), and a 2010 phosgene release at a DuPont plant in Belle, south of Charleston. (Phosgene gas, by the way, was used with widespread deadly effect as a chemical weapon in World War I.) Both incidents involved likely avoidable fatalities and injuries, not to mention potential longer-term environmental and health-related complications: the CSB reported that both accidents involved "regulatory deficiencies that were not identified or corrected through voluntary compliance or existing enforcement mechanisms prior to the incidents."[54] As a result of these two events, the CSB recommended in 2011 that the State of West Virginia take steps to develop new safety plans to prevent the unintentional release of hazardous chemicals in the future. The suggestions were rudimentary, such as requiring companies to submit new safety and prevention plans to the state and the public for review. But state officials never acted on the recommendations.[55] Had they, the Freedom spill could have been possibly avoided.[56]

Those living here who have knowledge of such incidents, past or present, are quick to point out that, in case after case, the regulations and oversight meant to protect the public seem to be at worst regularly ignored outright or at best only lackadaisically applied.[57] And they often cite the state's open antagonism toward regulation, especially environmental.[58] Indeed, the day before the spill incident, the first day of the West Virginia Legislature's 2014 regular session, the governor stated without equivocation in his State of the State address that "I will never back down from the EPA because of its misguided policies."[59] Though the governor's comments concerned coal in particular, and though he went on to say that the state could find ways to work with federal regulators, his expressed sentiment about the EPA—and its guidelines per the environment and public safety—articulate a familiar and oft-referenced position of many state and corporate leaders: that environmental regulation is ultimately a bad thing for West Virginia.[60]

This ostensible majority position seemed somewhat tempered with the passage of Senate Bill 373, the Aboveground Storage Tank Act (which the governor's office called the "Water Resources Protection Act"), signed into law by Governor Earl Ray Tomblin on the first day of April, only three months after the spill. The law—with bipartisan sponsorship and rapid approval—enacted several new measures to regulate aboveground chemical storage tanks and require multilevel disaster prevention plans like that originally recommended by the CSB several years before. Importantly, it required annual inspections of aboveground storage tanks to ensure they met basic standards for safe storage. It also required WVAW to gather information about hazardous chemicals stored near their facility, to develop "an early-warning monitor system," and to have plans in place to prepare for any future spill incident.[61] Though the bill's crafting was troubling at times (the governor, for example, organized an advisory group of business lawyers and industry lobbyists to help design the proposed legislation but excluded environmental and citizen groups), and though it left a broad range of problems and issues unaddressed and unresolved, many felt that its passage was a step in the right direction.[62]

That feeling was short-lived, however. The following year, a newly elected Republican-led legislature began to dismantle SB 373 and weaken its safe water regulations.[63] Ironically, the planned changes were in lockstep with much of the antiregulatory rhetoric of past majority Democratic-led legislatures and governorships (such as that expressed by Tomblin) for the eight-plus decades before the water crisis. Be that as it may, by midyear, the new 2015 legislature, responding in part to powerful lobbyists representing various business and industrial interests, passed SB 423 (signed into law in April 2015 by Governor Tomblin), which amended several major provisions lined out in SB 373. Thousands of aboveground storage tanks were exempted from the earlier regulation, and state inspections and other safety standards were relaxed. Remarkably, and paradoxically, it also provided provisions for limiting public knowledge of dangerous chemicals stored in tanks near or upstream of the treatment plant's raw water intake—including chemicals like MCHM—that could find their way into the potable water system.[64] Provisions of the new bill suggested that the Elk River spill could now plausibly unfold without the public ever knowing the identity of MCHM (or any other chemical) contaminating the drinking water system. In this strange turn of events, the new bill seemed more than just "a rollback of all we accomplished last year," as one legislative delegate put it.[65] To many, it turned what the year before had been called "the people's bill" on its head, making it even more difficult for the public to grasp, much less counter, the corporate interests and regulatory deficiencies that had facilitated the disaster in the first place.[66] The new law codified in SB 423 guaranteed that a replay of the Elk River spill would feature even greater confusion, uncertainty, and fear for the public, less transparency as to the dangers they faced, and greater anonymity for offending corporations such as Freedom Industries.

A little less than eighteen months after the spill, then, things seemed to be back to business as usual in West Virginia. And although the aboveground storage tanks that once stored MCHM and other chemicals had been demolished, deeper currents of distrust, uncertainty, and continuing lack of transparency kept the water crisis alive and unresolved. For the rest of the year and into the next, 2016, more studies emerged, raising more questions about long-term health effects; policy makers and regulators continued to push for additional rollbacks to water safety; and—among other things—more than a few people continued to report that they had serious misgivings about the safety of their water.[67] Indeed, the more Beth and I learned of the historical and current workings of the public water system here (not to mention the ongoing scientific studies of MCHM's toxicity), the less confident we felt about using our tap water for drinking or cooking.[68] Given West Virginia's well-documented track record, could we actually trust government or private interests to guard against another spill like this? And to make sure day in and day out that our water is safe?

The fact is "chemical leaks into the state's water still happen all the time," as local award-winning investigative reporter Ken Ward Jr. puts it: "Thousands of them are reported every year. More than 3,000 in 2013 alone, according to data from the West Virginia Department of Environmental Protection."[69] And, as many who live here can recount, often with multiple illustrative cases, offenders rarely face serious charges. Freedom Industries, for example, declared bankruptcy a little over a week after the spill, which put several lawsuits on hold and slowed claims against the company.[70] What's more, the criminal charges against Freedom officials were eventually reduced to misdemeanors (the case finally closed in February 2016), as part of a guilty plea agreement: the resulting sentences (ranging from three years' probation to thirty days in jail) and fines (less than $100,000 total) were considered exceedingly lenient by experts, especially given the magnitude of the original felony charges that Freedom officials faced (as well as other related charges, including obstruction of justice and federal charges of bankruptcy fraud).[71]

This is the kind of story ending that many West Virginia residents know all too well and, in many cases, have come to accept. Fortunately, though, the story doesn't end here. I could, of course, close by recounting the dozens of other federal and other legal cases that have materialized since the original spill and continue to evolve as I write this.[72] But that's quite beyond the task of this short background sketch and actually beside the point. The central point here is threefold: first, to note the historical context in which Sue Davis's comments materialized; second, to provide the social and cultural context that gives rise to the *dis*trust and uncertainty that frame Davis's comments; and third—and critically—to underscore the political context in which Davis's defiant tone still resonates with so many. Perhaps with this sociohistorical, cultural, and political context in mind, one can more easily grasp the gravity of her comments. Indeed, trust in water is arguably *the* defining issue in an evolving water crisis that still seems far from over.

Those issues of trust, or lack thereof, extend well beyond the uncertainty of the spill incident itself. In one of many studies conducted by various (and mostly out-of-state) organizations on the crisis's reverberating effects, researchers at the Harvard School of Public Health and the Georgetown University School of Nursing and Health Studies noted a year after the spill incident that "a lack of trust in public officials and the lingering odor of MCHM may explain why some residents refused to use the public water supply for months to a year after the water crisis was deemed over."[73] They point out that the "changing estimates, disparate statements, and uncertainty" that emerged around the spill event itself—"what seemed to be constantly changing facts" about MCHM and water safety—"undermined the public's trust in officials."[74] This was most certainly the case in the time during and after the spill. Indeed, Regina Lipscomb, an interviewee for this study, reported that she was still using bottled water for drinking and cooking in July 2017, fully three and a half years after the original spill, which cost her thirty dollars per month, adding a hefty 75 percent premium to her normal forty-dollars-per-month water bill.[75] Many others we interviewed reported similarly.

The Harvard and Georgetown study, however, glosses a deeper dimension of this trust: an ever-evolving sociohistorical, cultural, and political context that yields an a priori lack of trust, a *dis*trust that, for many here, is enlarged each time new uncertainties are introduced by yet another disaster event, in this case the water crisis.[76] For some, that a priori lack of trust is no doubt based on fear for their own well-being and that of family and friends; on misunderstandings over how scientific knowledge is produced, debated, and negotiated; or on confusing, conflicting, and disparate facts communicated by public officials at the time of the spill—all of which this study suggested.[77] Indeed, "I'm afraid of that water" was a refrain that we heard regularly and often. But clearly something more was at play. For many others, uncertainty and distrust don't just emerge from fear or confusion or lack of scientific knowledge. They also find their roots in *previous and ongoing experience* of living and working in Chemical Valley, remembering and voicing assorted stories of disaster in Appalachia, witnessing time and again the unconcealed privileging of corporate interests over those of citizens, and regularly dealing with polluters and other offenders who more often than not outmaneuver justice at the public's expense. Simply put, people here have many substantiated, defensible, and rational reasons for their misgivings about the safety of their tap water. Given all of this, though, these same people—everyday citizens like Sue Davis—have also found ways to mobilize around this event, voice their concerns, seek change, and actively push for a new kind of certainty when it comes to their water.

Which brings me to this story's middle, rather than its end.

## "And What Can *You* Do?"

After we moved to Charleston in 2005, Beth and I experienced other misadventures, such as the occasional (and unnerving) shelter-in-place drills and orders, which various chemical plant leaks, fires, and explosions seem to justify; and the derecho of 2012, which left many West Virginians without electricity for more than two weeks. But none of these was like the Elk River chemical spill, especially because the spill called into question for so many that which is so basic to survival. Nevertheless, on the morning after the spill incident, on January 10, we did what many others do during such times: unable to go to work or almost anywhere else, we went outside and started talking with our neighbors.

Within a few minutes, a neighbor relayed a story about a friend of his son's from Prenter, West Virginia. The boy had developed cancer at a young age, which, along with a range of other serious health problems in the boy's family and community, had been linked to the injection of coal waste into groundwater by Massey Energy (the same company held responsible in 2010 for the Upper Big Branch Mine disaster).[78] Because this groundwater fed the wells for Prenter residents, the health of many there, my neighbor conveyed, was forever changed. Their own well water, too, was now unusable. As a result, and as if to add insult to injury, Prenter had been recently added to the Charleston-based water system. My neighbor wondered aloud if Prenter residents could ever trust *any* water—much less any*body*—in West Virginia. "It sure makes me think about drinking my water," he said.

For the next several days, as we leaned on a range of friends living outside the affected area to take showers, wash clothes, and collect potable water, we heard a variety of stories like the one told by one of our health providers about the "cesspool of chemicals we live in here," or by a former chemical worker who eventually left his job at a local plant after being ordered to dump chemicals illegally, or by a local artist who had to be in a medical monitoring program for the rest of his life as a result of exposure to dioxins that affected thousands in his town, or by a coal engineer who had heard of the shady practices of a coal company competitor, or by one of our students, whose family regularly talked about and planned for the "next disaster." I initially thought these stories might be a way to voice empathy or concern or, alternatively, to navigate and make sense of the various and conflicting information unfolding about the spill. But as I heard story after story, I began to wonder if there might be something more to this.

Those wonderings hit home in a very personal way when I received a call from an out-of-state friend who had heard about the spill on the national news. He asked about our welfare and how things were going, of course, but our conversation soon turned to West Virginia. "It seems like West Virginia deserves this, don't you think?" he finally asked, and then reported that he had heard many others make similar statements. "I mean, you live in Chemical Valley. What do you expect?" I paused. I wasn't really sure what to say. "Why the hell do people continue to live there?" he continued after a short while. "Why the hell do *you* continue to live there?"

I admitted that the spill had prompted Beth and me to seriously consider leaving our jobs and West Virginia, a sentiment that only grew the more we learned about the overt abandon, corporate and governmental, that set the stage and practically invited the incident to happen. But for others, I said, based on what many had told us, connections to place and family loomed large. As a local business owner, Paul Gilmer Jr., put it: "There have always been many different things that happen [here] that make you think about, you know, 'Would I be better off in North Carolina? Or would I be better off in Ohio?' I always think about that. But there's nothing that's made me pack a suitcase. I'm committed here, my family's roots are down here. I've got my business roots down here."[79]

As important as are connections to place and family (a topic to which we return later in this book), something else also seems to keep many people here. And this, I think, really goes to the heart of the matter. Though West Virginians endure long traditions of outright negligence and abuse, they also live within streams of homegrown community action, among people who believe, and strongly so, that they can actually change how things are and make a difference (if only a small one) in their communities. The history of local activism in this state—and in the Appalachian region as a whole—is well known and much discussed among many who live here, but not so much among those outside the region, like my friend, who at the time lived and worked in New York City.[80]

This particular event, though, seemed to provoke new kinds of action, perhaps pressed by new levels of anger and frustration. As longtime activist and well-known West Virginia civil rights lawyer Paul Sheridan put it: "I feel like now I don't live in the world I thought I lived [in], and I thought I was pretty sophisticated about this stuff. I mean, I've been fighting different battles for many years and . . . I don't think of myself as somebody with great illusions about either the goodness of the corporations that bring us the goods and services we consume or the consistent thoroughness of all government investigators, many of whom I've worked alongside of in different ways. But this [water crisis] was like a new level of shock and disappointment."[81]

Consistent with a state- and region-wide tradition of community activism and nonprofit activity, during and after the spill, new groups began to surface (such as Advocates for a Safe Water System, discussed in chapter 8) and several existing organizations (such as the West Virginia Center on Budget and Policy, which helped sponsor the research behind this book in its earliest stages) marshaled citizens in new ways to act for change. To be sure, many people were radicalized to act in ways they hadn't before. One of these people was Jim Hatfield, a retired PhD chemical engineer, who, as he puts it, was discouraged by the "technical deficiency on the part of the water company and the State . . . [which] helped magnify the January 9 chemical spill into a nine-county, regional public drinking water disaster of national proportion. The event hurt many individuals and families economically, physically, and emotionally. Knowing this and then reading half-baked technical 'explanations' delivered by some officials in cavalier and arrogant tones propelled me into the land of activism."[82]

Jim Hatfield is important to mention here because he—along with his wife and partner, Trish Hatfield—provoked the original work that eventually led to this book. Here's the story. We had known Trish and Jim since soon after we moved to West Virginia. At the time, Trish was finishing her degree in the graduate humanities program I direct (today, she serves as the unit's program assistant). An activist in her own right involved in several nonprofit and other community groups and owner of a consulting firm based on appreciative inquiry, Trish immediately introduced us to a wide range of community change agents throughout the state and region and eventually helped to initiate several of our program's projects, including the West Virginia Activist Archive.[83] In the days immediately following the spill, it became clear that Jim in particular was growing more and more frustrated by how technical inconsistencies and water company and government officials' associated "information spinning" (such as this being an "isolated event") had nurtured and continued to feed an emerging, and serious, water crisis.

So it was that early one morning I got a call from Trish: "Jim wants to talk to you. Can we have lunch?" I agreed. We met in early April 2014 at a local restaurant, Blues BBQ—owned and operated by another one of our graduate students, Jay Thomas (author of chapter 5). Though the spill incident was already three months old, the still-evolving water crisis remained very much on people's minds: new information surfaced weekly about Freedom Industries, for example, and restaurants, like Jay's, still served bottled water to their customers. So, in sum: though I wasn't quite sure about the specific topic of our meeting, I suspected it might have something to do with Jim's concern about the spill's still-rumbling effects.

I was right. But Jim had recently become focused—even obsessed—with finding an academic institution or a group that could do a more extensive impact survey beyond a local but widely circulated economic impact estimate

carried out soon after the spill, which, as the report noted, was very preliminary in its findings and short on conveying longer-term impacts.[84] Jim, who
is incredibly meticulous and detailed while at the same time deeply curious
and always refreshingly thoughtful, had grown frustrated as he sought out
social scientists to help and had encountered several roadblocks as he initiated
contacts at various academic institutions locally, regionally, and nationally to
make his case. He introduced some questions about how best to proceed: What
academic units, for example, and which universities would be best positioned
to carry out such a study? What other kinds of studies could be done? Could
such studies ascertain other kinds of impact, such as social or cultural? Who
would do them? Which turned Jim to a new set of questions—directed at me.

"And what can *you* do?" he finally asked. How could I, Jim wanted to know,
contribute to a larger study of impact? How could the graduate humanities
program and our various community-university research partnerships help?
And how might we include other faculty—or students, like Jay—in making a
concerted change here? What were the possibilities for research? How could my
focus and expertise be of service here? Had I considered doing a study myself?

Yes, I had, I reported. But, I also admitted, I was so angry and frustrated
about this whole mess that any thought of doing research on the crisis had
sunk into the background. Too, I just didn't feel like I could do it right then.
I was incredibly busy, I said, working on multiple other research and writing
projects, and as a full-time professor and university administrator, my plate,
to say the least, was (and always is) overflowing. I made my case forcefully,
then offered some ideas about which universities, disciplines, departments,
and individual researchers might be able to lend a hand.

Jim accepted my various excuses, for the time being. We turned to sharing
various stories we had heard. Both Jim and Trish—who live in Saint Albans,
which was outside the affected zone, and so had hosted many folks looking
for potable water, showers, and the chance to wash clothes—talked about
stories they had heard and about the new group in which Jim was now involved, Advocates for a Safe Water System. After a while, I shared some of
the stories I had heard and how different they seemed to be from what I had
read in preliminary public health studies, magazines, newspapers, and other
media, which overwhelmingly focused on how residents continued to avoid
their water, the studies suggested, because of irrational fears or confusion over
disparate scientific findings. I told Jim and Trish that I felt as if these stories
and the voices in them were not being heard, that these down-on-the-ground,
day-to-day experiences were not being factored into larger understandings of
the spill's impact. "Now you're talking like an ethnographer," Trish interjected.

"Seems like a topic for some sort of ethnographic study," Jim added, returning to his earlier point. Both Trish and Jim knew ethnography well: Trish used ethnographic approaches in her final graduate project, and both Jim and Trish had read, out of sheer curiosity, several of my previous books on anthropology and ethnography. So they both knew ethnography's potential for engaging local story and dialogue. We invariably turned to talking, again, about various possibilities for a study of some sort. But this time around I was warming up to the idea, and all the other things going on in my life just didn't seem as important at that moment. Indeed, the Hatfields were working their magic: a magic I had encountered before, which involved in one way or another getting me—or Beth, other faculty, students, our program—included in something local and meaningful, something important. For the first time since the crisis, my anger and frustration began to give way to more positive feelings, thanks to the Hatfields. So we left lunch that day with the agreement that we would work on tracking down some sort of support to at least get us started on a preliminary study, which would help to document the experience of living through, and then navigating, our ongoing water crisis. Perhaps, on some level, we could contribute to larger discussions of sociocultural and economic impact.

## Collecting Oral Histories:
## A Brief Note on Methods and Approach

Jim, Trish, and I eventually turned to the West Virginia Center on Budget and Policy for help. Under their auspices, we secured support from the Oral History Association's Emerging Crises Oral History Research Fund to carry out oral history research of the Elk River spill.[85] This section briefly summarizes the methods we used to collect these oral histories.[86] In chapter 3, I return to a more in-depth discussion on how this preliminary research morphed into collaborative ethnography, which came to include a much larger group including various community members, faculty, students, and anthropologist Brian Hoey, author of chapter 2 and the interlude.

Within a few months after our lunch, Jim, Trish, and I had assembled a small research team. With intentions of establishing a more long-term research project that could potentially involve multiple partners and expand into other realms of research activity over time (such as collaborative ethnography), we asked Beth to join us, then assembled a four-member research team made up of folks located in and around the Charleston area to serve as the project's oral history researchers. These included Jim and Trish, Cat Pleska (well-known

West Virginia author and associate graduate faculty in our program), and Marla Griffith (an independent researcher with extensive historical research experience), all of whom possessed wide-ranging connections to various persons affected by the spill. Around the same time, I also began talking with Brian Hoey, my anthropology colleague based in Marshall University's department of sociology and anthropology, and we began exploring ways we might work together to address the spill and its aftermath.

The oral history team first decided that our foremost responsibility should be to describe people's experience of the spill, whatever that might be. We recognized at the outset, of course, that it would be impossible to expect our study to be fully representative of the water crisis. We instead sought to identify an appropriate cross-section of individuals who had experienced the spill to varying degrees. Though few in the area were untouched by the crisis, we assumed that experience would vary in degree based on differences related to factors like geographical location (rural or urban), ethnic or racial background, class, work, age, religion, family status, level of political involvement, and so on. We thus focused on identifying categories of experience that might also help us to identify and then describe a broad cross-section of experiences associated with the spill. These categories included those such as single mothers, seniors, white-collar professionals, emergency responders, and low-wage workers; Anglo, African, or Latin Americans; people living in Charleston and in rural areas like neighboring Lincoln County; and so forth. Such categories, of course, were in and of themselves abstractions (many blended into one another) and thus were incomplete in their scope. But they did provide, in part, the base for an emergent research design upon which our team could (1) index the cross-section of individuals we sought to interview; (2) make choices about which team member would interview which individuals (and prevent unintended overlap or saturation of one group at the expense of others); and, perhaps most importantly, (3) consciously and regularly attend to issues of diversity within and across interviews (which we had via regular email exchanges and periodic meetings).

In a series of initial meetings, we put together a list of research questions to use as the project's interview guide. Given the range of oral history interviews we planned, we developed twenty-six diverse questions (grouped into broad categories), which project members could pull from in interviews based on whom they were interviewing. Most pulled about ten to twelve to initiate a given interview. As the project unfolded, we added and subtracted questions from this list, of course, and developed the larger project's interview guide further as our understandings became more nuanced. In addition to collecting information about background, work, and family (e.g., "Tell us something about who you are"), we sought to use these evolving questions to gauge various, overlapping aspects of the spill experience itself, including, for instance

- initial encounter ("Could you describe when you first suspected or knew something was wrong?")
- feelings immediate and over time ("Can you describe some of the feelings you were having at the time?" or "How did your feelings about the spill change during the crisis?")
- life adjustments ("Did your personal/work/school life change as a result of the spill/crisis?" or "Could you describe the daily routine you developed to deal with the spill/crisis?")
- physical conditions ("Did you ever get sick? If so, could you describe what your symptoms were?")
- short- and long-term effects ("How has this affected your life? Immediately, a few days after the spill, a month later, the present, the future?")
- shifts in perception ("Has this made you think differently about your water?" or "Has this made you think differently about your political leaders?" or "Has it changed how you think about state or federal regulations?")
- dealing with ongoing and future environmental crisis issues ("What do you think we need to do to address this continuing situation?" or "For you, what is safe drinking water? What is a safe environment?")

During the initial research period, which lasted until December 2014, the team conducted thirty-six interviews (though a few interviews continued after that period). Each interview was digitally recorded and archived in a central location. For each interview, research team members also provided detailed logs (multipage, in-depth summaries of interview content at five- to ten-minute intervals) and identified interview sections as well as entire interviews that should be slated for transcription.

The original recordings, detailed logs, transcriptions, and, importantly, ongoing team discussions began to yield some preliminary findings as the research progressed. All participants did not encounter the event the same way, of course, but collected narrations did seem to provide both a sufficient baseline and, on some levels, an appropriate cross-section of experiences. As expected, commonalities surfaced between all the interviews—such as the process of coming to know about the spill and the "do not use" water order (many heard via various media accounts, from friends, or from an official phone call from West Virginia American Water)—but the particulars were also wide ranging. Marla Griffith's interview with, for example, local physician Shelda Martin was very different from the interview Trish Hatfield carried out, for instance, with the client of a local homeless shelter or Cat Pleska's interview with a local artist or Jim Hatfield's with civil rights lawyer Paul Sheridan.

I was somewhat surprised to find that a few of the interviewees narrated their experiences in terms very much like those reported in various media: that their continued fear of the water seemed, even to themselves, irrational, a purely psychological response. In a similar vein, unsurprisingly, we also collected wide-ranging descriptions of various physical responses, such as bodily reactions to consuming contaminated water or even coming into contact with the water system post-spill. Cat Pleska's interview with Carla McClure—about how she and her husband, Phillip, became ill during the flushing process—is an example.

Though experiences such as these were also regularly documented in local media accounts, more than a few residents expressed indifference toward the water crisis and thought governmental leaders, public health officials, reporters, activists, and others (including researchers like us) were making more of the event than necessary. Jim Hatfield, for example, interviewed local writer Ginger Caudill, also disabled and homebound, who described how she watched the event unfold on social media: "I read, at the time, all the Facebook posts . . . and I would follow the general scuttlebutt of what people were saying. And it's shocking how many people really did not think there was any problem. It's shocking how many people think that this whole thing has been a bunch of hooey, that none of it was real—that they just got everybody worked up for no reason—that they continued to use the water and never had any problem. And they don't even believe that a problem has ever existed."[87]

Given the expressed doubts about the seriousness of the event, issues of trust and distrust (with facilities, state and federal leadership, media, and industry) regularly surfaced in many if not most interviews. Indeed, a majority of our interviewees described a process of questioning their previous levels of trust. Sharon Moriarty, for example, an employee of an in-home care service for seniors interviewed by Marla, described her own process as the spill event unfolded: "I became more frustrated, and I felt like, how could this have gone on without being checked? The last safety check [of the chemical storage tanks] had been what? Twenty-five years prior? Why weren't there more consistent checks for that setup? I mean, it's a chemical . . . and it's near our water source, so why wasn't it monitored more closely? And why wasn't it safer, more contained? It was amazing to me that it had gone unchecked for so long."[88] Importantly, for several narrators, these expressions were often framed by how previous experiences helped to shape current experiences with and reactions to the spill.

This finding, as is perhaps now obvious, came to powerfully inform my own understanding of the event and how I chose to frame this introductory chapter. Along these same lines, issues of short- and long-term costs (including those financial, such as the cost of a suddenly idle workforce or the cost of bottled water, and symbolic, such as to the state's reputation) surfaced in more than a few interviews. One of Beth's interviewees, Rebecca Roth, a young professional living and working in Charleston, mentioned how the spill and ensuing water crisis had shaken her own and others' views of the area, including its long-term health safety (part of this interview is included in the introduction).

Through the process of carrying out this oral history research, I, for one, came to understand better the struggles many of the interviewees expressed, to feel more deeply what it meant, for instance, for many West Virginians to make painful choices as they considered staying or leaving West Virginia. It was, indeed, complicated. As local African American business owner Paul Gilmer Jr. said to Trish Hatfield: "I've got my business roots down here."[89] The same could also be said, of course, of other roots of family and place. I mention this issue again here because these stories led us, the oral history research team, to deeper understandings of how the event provoked larger numbers of people to consider more directly the issues of water quality, environment, regulatory agencies, and their elected representatives. In many cases, interviewees surprised us with their persistent hope in the face of what otherwise seemed a hopeless situation; most often, their hope was for much-needed change in corporate, governmental, and community consciousness that is, indeed, still emergent and still evolving. As one citizen activist, Becky Park, put it in an interview with Trish Hatfield: "I am very happy with the fact that we have a lot more people paying attention to all of those issues."[90]

These core and critical issues of trust and distrust; of commitments to family, place, and business; of deciding to stay or leave; of getting involved, even becoming radicalized, or not; among other things, are explored in the book's remaining chapters, coauthored by Cat Pleska and Trish and Jim Hatfield, members of the oral history research team, and by others as well. The inclusion of those others as coauthors requires further explanation and points us in the direction of how this project expanded into collaborative ethnography.

But before I launch into how this project developed as collaborative ethnography, a deeper discussion of human disaster is called for. Brian A. Hoey introduces this discussion in chapter 2. I return to our project and collaborative ethnography in chapter 3.

## Notes

1. Sue Davis, quoted in Ken Ward Jr., "CSB Agrees to Report 'Addendum,'" *Charleston Gazette-Mail*, September 29, 2016, 8A.
2. See U.S. Chemical Safety and Hazard Investigation Board, Investigation Report: Chemical Spill Contaminates Public Water Supply in Charleston, West Virginia, Report No. 2014-01-I-WV (Washington, D.C.: U.S. Chemical Safety and Hazard Investigation Board, 2016).
3. Though MCHM became common parlance for the contaminant, residents were later informed that the contaminant, in fact, was a mixture of isomers of 4-MCHM together with a flotation chemical referred to as PPH, a propylene glycol phenyl ether. The precise composition of the contaminating mixture is important, of course, mainly because failure to recognize this led to toxicological studies being performed on only one of the several components. In this book, the term *MCHM* will be used for convenience, although it is understood to represent a more complex chemical mixture including MCHM isomers and PPH. For more on this, see CSB, Investigation Report, esp. p. 19.
4. Evan Osnos, "Chemical Valley: The Coal Industry, the Politicians, and the Big Spill," *New Yorker*, April 7, 2014, accessed October 15, 2016, http://www.newyorker.com /magazine/2014/04/07/chemical-valley.
5. CSB, Investigation Report, see p. 111ff.
6. Osnos, "Chemical Valley."
7. Ken Ward Jr., "Just 36% Drinking Tap Water since MCHM Leak," *Charleston Gazette-Mail*, May 12, 2014. See also David Latif, Rahul Gupta, and Elna Savoia, "Communications during the West Virginia Water Crisis: A Survey of the Population" (Charleston: Kanawha-Charleston Health Department, 2014).
8. See, e.g., David Gutman, "'WaterFest' Generates Goodwill, but Concerns Linger," *Charleston Gazette-Mail*, August 9, 2014.
9. See, e.g., "In West Virginia, Fear about Safety of Drinking Water Persists," National Public Radio, December 27, 2014.
10. Gregory Button, *Disaster Culture: Knowledge and Uncertainty in the Wake of Human and Environmental Catastrophe* (Walnut Creek, CA: Left Coast Press, 2010), esp. chapter 7.
11. Button, 14–15.
12. Button, see esp. chapter 9.
13. Button, 15.
14. The following brief description is not meant to be an extensive or exhaustive survey. I offer it here only as a brief contextualization of the oral histories in subsequent chapters. More in-depth description and analysis can be found in a number of sources, many of which are cited in this chapter. In particular, my timeline of events draws from my own firsthand experience, from the CSB's Investigation Report, and from the investigative reporting of *Charleston Gazette-Mail*, which has covered and continues to cover the spill and its accompanying water crisis into the present. Its reporting includes regular and constant reference to scientific and other studies

about the spill, many of which are referenced below and inform my discussion, too. A good resource for following, in particular, the timeline of local reporting from January 9, 2014, to the present can be accessed online at *Charleston Gazette-Mail's* "Water Crisis" page at https://www.wvgazettemail.com/news/special_reports /water_crisis/.

15. This is, of course, an incredibly abbreviated description of the chemical industry in and around Charleston, and I am in no way an expert on its history. For a more in-depth survey, see, e.g., Nathan Cantrell's "West Virginia's Chemical Industry," *West Virginia Historical Society Quarterly* 18, no. 2 (2004): 1–15. More about slavery in Charleston and its relationship to the salt industry can be found in R. Eugene Harper's "Slavery in Charleston, 1830–1860," in *Constellations: An Anthology of the Marshall University Graduate Humanities Program, Celebrating Thirty Years (1979– 2009)*, ed. Kathryn Santiago (South Charleston: Occasional Publications of the Graduate Humanities Program, Marshall University, 2009), 39–47.

16. For a more detailed summary of the day's events, see CSB, Investigation Report, 7–9.

17. Charlotte Ferrell Smith, "Water Warning Now in 9 Counties; Emergency Supplies on Order," *Charleston Gazette-Mail*, January 9, 2014.

18. Shelda Martin, interview with Marla Griffith, August 26, 2014. Compare, e.g., with Luke E. Lassiter, *The Power of Kiowa Song* (Tucson: University of Arizona Press, 1998), 231–32.

19. CSB, Investigation Report, 1.

20. As one might expect, reports of price-gouging surfaced during this time, though they seemed to be isolated events. See, e.g., Lydia Nuzum, "Putman Stores Price-Gouged Water Buyers, Morrisey Says," *Charleston Gazette-Mail*, February 14, 2014.

21. Local pastor and activist Matthew Watts reported, for example, that "the initial places for the water distribution were such that if you didn't have transportation it would be hard for you to get water because you could only carry it so many blocks. We believe that many people in this community probably did use the water when they were advised not to use it, because they just could not get enough water." See Gabe Schwartzman's chapter 9.

22. See, e.g., Rick Steelhammer, "Southern W.Va. Scrambles for Water," *Charleston Gazette-Mail*, January 10, 2014; Staff Reports, "St. Albans Man Arrested following Incident at Water Stop," *Charleston Gazette-Mail*, January 13, 2014; Ken Ward Jr., "State Acknowledges It Had No Plan for Freedom Spill," *Charleston Gazette-Mail*, January 14, 2014.

23. Ward, "State Acknowledges."

24. CSB, Investigation Report, 53–54.

25. CSB, 53.

26. See Ken Ward Jr., "100,000 May Have Felt Symptoms from MCHM," *Charleston Gazette-Mail*, April 22, 2014.

27. Marshall University Center for Business and Economic Research, "CBER Calculates Impact from Chemical Spill into Elk River," 2 February 2014.

28. See Rachel Molenda, "Restaurants, Bars in Chemical Leak Area Start to Reopen," *Charleston Gazette-Mail*, January 12, 2014; Lydia Nuzum, "Local Initiatives Aim to Help Restaurant, Service Workers," *Charleston Gazette-Mail*, January 13, 2014.

29. Nuzum, "Local Initiatives"; Jared Hunt, "Senator Encourages Others to 'Turn Up the Tips,'" *Charleston Gazette-Mail*, January 13, 2014.

30. See Caitlin Cook, "Back to Business but not Tap Water," *Charleston Gazette-Mail*, January 16, 2014.
31. Jonathan Matisse, "In Charleston, Business Interruption Losses from Chemical Spill Take Toll," *Insurance Journal*, May 12, 2014.
32. CSB, Investigation Report, 53–54.
33. See, e.g., Ken Ward Jr., "How Do They Know Water's Safe at 1 ppm?" *Charleston Gazette-Mail*, January 13, 2014; Ken Ward Jr., "Experts: Not Enough Data to Judge MCHM Dangers," *Charleston Gazette-Mail*, January 18, 2014.
34. Richard Denison, Environmental Defense Fund, quoted in Ward, "How Do They Know."
35. See CSB, Investigation Report, 117. See also Mays Mackenzie, "Several Zones Given Go-Ahead to Start Flushing," *Charleston Gazette-Mail*, January 13, 2014; Rusty Marks, "Water Company Gives Instructions on Flushing Process," *Charleston Gazette-Mail*, January 13, 2014; West Virginia American Water, "How to Flush Your Plumbing System," accessed November 17, 2016, http://www.amwater.com/files /WV%20-%20How%20to%20flush.pdf.
36. CSB, Investigation Report, 117.
37. CSB, 2, 9; Ken Ward Jr., "MCHM 'Flushing' Likely Exceeded EPA Health Guidance, Study Says," *Charleston Gazette-Mail*, July 16, 2016.
38. Carla Thomas McClure, interview with Cat Pleska, July 10, 2014.
39. See Ken Ward Jr., "Pregnant? Bottled Water Recommended," *Charleston Gazette-Mail*, January 15, 2014. But see, too, Ken Ward Jr., "CDC: Pregnant Women Should Have Been Warned about Water Sooner," *Charleston Gazette-Mail*, January 22, 2014.
40. CSB, Investigation Report, 19.
41. See, e.g., Ken Ward Jr., "Information on Leak's 2nd Chemical 'Very Limited,'" *Charleston Gazette-Mail*, January 22, 2014; Ken Ward Jr., "Elk River Leak Included Another Chemical," *Charleston Gazette-Mail*, January 21, 2014; and Ken Ward Jr., "CDC Tested Only 1 Ingredient of Elk River Chemical," *Charleston Gazette-Mail*, January 18, 2014.
42. CSB, Investigation Report, 55.
43. Ken Ward Jr., "Experts Warn 'Flushing' Might Not Have Worked," *Charleston Gazette-Mail*, January 29, 2014. See also Lori Kersey, "Some Don't Trust the Water Yet," *Charleston Gazette-Mail*, January 14, 2014; but see Ken Ward Jr., "Pipe Tests Show 'Limited' MCHM Permeation," *Charleston Gazette-Mail*, June 10, 2014.
44. CSB, Investigation Report, 55–56. See also Ward, "MCHM 'Flushing' Likely Exceeded."
45. See CSB, Investigation Report, 60–61. See also Ken Ward Jr., "MCHM Odors Show Chemical Not Gone, Experts Say," *Charleston Gazette-Mail*, March 18, 2014.
46. See Ken Ward Jr., "Tap Water Sampling Finds MCHM in Some Homes," *Charleston Gazette-Mail*, February 7, 2014; Ken Ward Jr., "Home Testing a Start, but Questions Remain," *Charleston Gazette-Mail*, February 6, 2014; Mackenzie Mays, "Two More Kanawha Schools to Re-flush Pipes," *Charleston Gazette-Mail*, February 4, 2014; Mackenzie Mays, "Water Response Team Called to Four Schools Monday," *Charleston Gazette-Mail*, February 17, 2014; Ken Ward Jr., "Expert: Flushing Effectiveness 'Mixed,'" *Charleston Gazette-Mail*, March 14, 2014; Ken Ward Jr., "Many Chemical Spill Questions Remain, First WVTAP Report Say," *Charleston Gazette-Mail*, March 17, 2014; Ken Ward Jr., "MCHM Could Be More

Toxic Than Reported, New Study Says," *Charleston Gazette-Mail*, July 10, 2014; Ken Ward Jr., "Study Warns of MCHM Toxicity," *Charleston Gazette-Mail*, May 12, 2015.

47. See, e.g., David Gutman, "Public Health Officials Differ on Water Use," *Charleston Gazette-Mail*, February 6, 2014. Brian Hoey is responsible for encouraging our research team to think about the flushing and its aftermath as a second contamination experience.

48. See, e.g., Ken Ward Jr., "Testing Shows Chemical Decline, but Questions Persist," *Charleston Gazette-Mail*, January 23, 2014; Rachel Molenda, "Water Questions Unanswered in Rural W.Va.," *Charleston Gazette-Mail*, January 22, 2014.

49. See, e.g., David Gutman, "Freedom Executive Kennedy Had Felonies," *Charleston Gazette-Mail*, January 12, 2014.

50. See CSB, Investigation Report, 35–36; Ken Ward Jr., "Freedom Industries Cited for Elk River Spill," *Charleston Gazette-Mail*, January 10, 2014.

51. See, Ward, "Freedom Industries Cited"; and David Gutman, "Freedom Industries Execs Are Longtime Colleagues," *Charleston Gazette-Mail*, January 10, 2014.

52. See CSB, Investigation Report, 11, 35–36. See also Ken Ward Jr., "Feds: Freedom Knew about Problems for Years," *Charleston Gazette-Mail*, January 8, 2015.

53. Much has been written on the Buffalo Creek Flood disaster. See, e.g., Kai T. Erikson, *Everything in Its Path: Destruction of Community in the Buffalo Creek Flood* (New York: Simon and Schuster, 1978).

54. CSB, Investigation Report, 61.

55. See Ken Ward Jr., "State Ignored Plan for Tougher Chemical Oversight," *Charleston Gazette-Mail*, January 12, 2014.

56. For example: Even under existing required regulations, at least two of Freedom's pollution prevention documents were never submitted or reviewed by state officials for Freedom's water pollution permit when the company took control of the site. See Ken Ward Jr., "DEP Never Saw Freedom's Pollution Control Plans," *Charleston Gazette-Mail*, February 1, 2014. Ward reported that "under a DEP-approved water pollution permit for the site, Freedom Industries was required to prepare a storm-water pollution prevention plan and a groundwater protection plan. Neither plan was among the documents contained in Freedom's permit files at the DEP's Water and Waste Management office." See also Ken Ward Jr., "Why Wasn't There a Plan?," *Charleston Gazette-Mail*, January 12, 2014.

57. See, e.g., the oral histories that follow in this book's subsequent chapters.

58. See, e.g., Rob Byers, "Will the Water Crisis Change Us for the Better?," *Charleston Gazette-Mail*, May 8, 2014.

59. Governor Earl Ray Tomblin, State of the State Address, January 8, 2016, Office of the Governor, accessed November 1, 2016, http://www.governor.wv.gov/media/pressreleases/2014/Pages/GOVERNOR-DELIVERS-STATE-OF-THE-STATE-ADDRESS.aspx.

60. See, e.g., David McKinley, "Connecting the Dots on Regulations," *Charleston Gazette-Mail*, June 2, 2015, which expresses this sentiment adamantly from the viewpoint of a state legislator.

61. See Eric Eyre, "Tomblin Signs Storage Tank Bill," *Charleston Gazette-Mail*, April 1, 2014. See also "Governor Tomblin Signs SB 373, the Water Resources Protection Act," Office of the Governor, accessed November 2, 2016, http://www.governor.wv.gov/media/pressreleases/2014/Pages/Governor-Tomblin-Signs-SB-373,-The-Water-Resources-Protection-Act.aspx; and, for the full text of the bill, Senate Bill

No. 373, West Virginia State Legislature, accessed November 2, 2016, http://www
.wvlegislature.gov/bill_status/bills_text.cfm?billdoc=SB373+SUB2+ENR.htm&yr=
2014&sesstype=RS&i=373.

62. See Ken Ward Jr., "Tomblin Meeting on Chemical Tank Bill Excluded
Environmentalists," *Charleston Gazette-Mail*, February 4, 2014; and David Gutman,
"After Passage, Steps Remain for Tank Bill," *Charleston Gazette-Mail*, March 9,
2014.

63. Ken Ward Jr., "On MCHM Leak Anniversary, Groups Get Set for Push to Relax
Water Protections," *Charleston Gazette-Mail*, January 9, 2015; Ken Ward Jr., "A
Year Later: Chemical Spills Continue, and So Do Questions," *Charleston Gazette-
Mail*, January 8, 2015; and Ken Ward Jr., "Changes Could Leave Thousands of
Chemical Tanks out of Regulations," *Charleston Gazette-Mail*, January 15, 2015.

64. Ken Ward Jr., "Chemical Tank 'Rollback' Appears Headed for Passage," *Charleston
Gazette-Mail*, March 10, 2015. See also "Governor Tomblin Signs SB 423,
Amending the Aboveground Storage Tank Act," Office of the Governor, accessed
November 3, 2016, http://www.governor.wv.gov/media/pressreleases/2015/Pages
/GOVERNOR-TOMBLIN-SIGNS-SB-423,-AMENDING-THE-ABOVEGROUND
-STORAGE-TANK-ACT.aspx; and, for the full text of the bill, Senate Bill No. 423,
West Virginia State Legislature, accessed November 3, 2016, http://www.legis
.state.wv.us/Bill_Status/bills_text.cfm?billdoc=SB423%20SUB1%20enr.htm&yr
=2015&sesstype=RS&i=423.

65. Ward, "Chemical Tank 'Rollback.'"

66. Eyre, "Tomblin Signs Storage Tank Bill."

67. See, e.g., Ken Ward Jr., "Latest MCHM Test Results Raise More Questions,"
*Charleston Gazette-Mail*, August 5, 2016; Ken Ward Jr., "Final Federal MCHM Study
Leaves Same Questions Unanswered," *Charleston Gazette-Mail*, July 8, 2016; Ken
Ward Jr., "DEP Water Rule Proposal Could Increase Cancer-Causing Discharges,"
*Charleston Gazette-Mail*, August 6, 2016; Ken Ward Jr., "WV's controversial
Water-Quality Rule Change Stays," *Charleston Gazette-Mail*, August 26, 2016; Ken
Ward Jr., "Citizens Urge DEP to Abandon Water Rule Changes," *Charleston
Gazette-Mail*, August 9, 2016.

68. See Ward, "Study Warns of MCHM Toxicity," for a review of several scientific
studies. The online version of the article (at www.wvgazettemail.com) includes
links to those studies.

69. Ken Ward Jr., "Unanswered Questions Remain: Water Crisis, One Year Later,"
*Charleston Gazette-Mail*, January 9, 2015, 5A.

70. See, e.g., Kate White, "Freedom Industries Files for Bankruptcy," *Charleston
Gazette-Mail*, January 18, 2016.

71. See Ken Ward Jr., "Freedom Officials Plead Guilty in Leak Probe," *Charleston
Gazette-Mail*, March 16, 2015; Ken Ward Jr., "Fraud, Obstruction at Issue in Gary
Southern Sentencing," *Charleston Gazette-Mail*, February 16, 2016; and Ken
Ward Jr., "Freedom Spill Sentences Too Lenient, Experts Say," *Charleston Gazette-
Mail*, February 21, 2016. See also "Appendix B: Freedom Officials," in CSB,
Investigation Report, 118, for a six-person breakdown of charges, sentencing, and
fines.

72. For example, the residents of Charleston and the surrounding area are, as I write
this, digesting terms of a tentative class action settlement filed against West
Virginia American Water and the Eastman Chemical Company. See, e.g., Ken

Ward Jr., "How Might the MCHM Deal Affect You?," *Charleston Gazette-Mail*, November 2, 2016.

73. Michael A. Stoto, Rachael Piltch-Loeb, and Elena Savoia, "The Public Health System Response to the 2014 West Virginia Water Crisis" (Cambridge, MA; Washington, D.C.: Harvard School of Public Health and the Georgetown University School of Nursing and Health Studies, 2015), 7.

74. Stoto et al., 8, 9.

75. Reginia S. Lipscomb, personal communication with Jim Hatfield, July 14, 2017.

76. See, e.g., Andrew J. Whelton, LaKia McMillan, Matt Connell, Keven M. Kelley, Jeff P. Gill, Kevin D. White, Rahul Gupta, Rajarshi Dey, and Caroline Novy, "Residential Tap Water Contamination following the Freedom Industries Chemical Spill: Perceptions, Water Quality, and Health Impacts," *Environmental Science and Technology* 49 (2015): 813–23.

77. See Stoto et al., "Public Health System Response."

78. Much has been written about the Prenter case. For a brief synopsis, see, e.g., Robert Johnson, "Everyone Who Uses Coal Power Should See What's Happening to This West Virginia Mining Town," *Business Insider*, February 4, 2014, accessed November 29, 2016, http://www.businessinsider.com/prenter-hollow-west-virginia-faces -contamination-2014-1. An in-depth interview with Prenter residents and their experience fighting Massey Energy can be accessed on WVWaterHistory.com, a site developed by Gabe Schwartzman and Alicia Willett. See chapter 9 for further description by Schwartzman.

79. Paul Gilmer Jr., interview with Trish Hatfield, September 11, 2014.

80. We touch on this history briefly and its impact in the subsequent chapters. But for more in-depth description of change movements in West Virginia and the larger Appalachian region, see, e.g., Stephen L. Fisher, Patti Page Church, Christine Weiss Daugherty, Bennett M. Judkins, and Shaunna L. Scott, "The Politics of Change in Appalachia," in *A Handbook to Appalachia*, ed. Grace Toney Edwards, JoAnn Aust Asbury, and Ricky Cox (Knoxville: University of Tennessee Press, 2006), 85–100. See also Stephen L. Fisher and Barbara Ellen Smith, eds., *Transforming Places: Lessons from Appalachia* (Urbana: University of Illinois Press, 2012). For more on local activists involved in this and other issues in West Virginia, see, e.g., our ongoing oral history project on activism in West Virginia, the West Virginia Activist Archive, online at http://www.marshall.edu/graduatehumanities /wvactivists/.

81. Paul Sheridan, interview with Jim Hatfield, August 26, 2014; revised, personal communication with Jim Hatfield, August 23, 2017.

82. Jim Hatfield, quote excerpted from chapter 8 herein.

83. See West Virginia Activist Archive, http://www.marshall.edu/graduatehumanities /wvactivists/. In *Doing Ethnography Today*, Beth and I describe an earlier iteration of this project, which was initiated by Trish Hatfield. See Elizabeth Campbell and Luke Eric Lassiter, *Doing Ethnography Today* (Oxford: Wiley-Blackwell, 2015), 30–32.

84. See Marshall University Center for Business and Economic Research, "CBER Calculates Impact."

85. The West Virginia Center on Budget and Policy is one of our program's community partners. For more on this, see Water Crisis Oral History, http://www.marshall .edu/graduatehumanities/water-crisis-oral-history/.

86. Parts of what follows pull directly from Luke Eric Lassiter, "Oral Histories of the

Charleston WV Chemical Spill," final report submitted to the Oral History Association's Emerging Crises Oral History Research Fund (Charleston: West Virginia Center on Budget and Policy, 2015).

87. Ginger Caudill, interview with Jim Hatfield, July 22, 2014.
88. Sharon Moriarty, interview with Marla Griffith, July 12, 2014.
89. Gilmer, interview with Trish Hatfield, September 11, 2014.
90. Rebecca Park, interviewed by Trish Hatfield, September 10, 2014.

# CHAPTER 2

# Exploring the (Human) Nature of Disaster: Meaning and Context

Brian A. Hoey

## Background

I don't watch much network television, but I did have an urge to tune in to my local news to keep me company while home alone that fateful January evening, shortly before the start of our spring semester at Marshall University. What I got, instead of the usual chatter, was the governor announcing that there had been a chemical release near the water intake for the capital city and a total of nine counties. He informed the public that the drinking water had been contaminated as a result of a leak at a chemical storage facility. This was no inane talk—at least not yet. Such inanity would follow in the days and weeks to come, as evidenced by press conference statements by the governor himself, who stated at one point that "this is not a coal company incident, this was a chemical company incident" and added at another that "as far as I know, there was no coal company within miles."[1] These were attempts by the governor to put at least some rhetorical distance between a chemical "spill," as most would come to refer to what happened at the Freedom Industries site, and the coal industry. He appeared committed to establishing a narrative that the release of several thousand gallons of chemical stored specifically for and used exclusively by the coal industry as a "flotation reagent" in the refinement of coal was not directly connected to that industry's physical, historical, and ideological prominence in the state.

In my experience, it was exceptional to have a head of state telling hundreds of thousands of his citizens not to use their tap water except to flush toilets. Since that time, however, we've had widely reported lead contamination of drinking water affect tens of thousands of people in the cities of Flint, Michigan, and Newark, New Jersey. Both crises have required dramatic announcements as well as public reckonings of elected officials. Despite my shocked reaction to the sheer gravity of the West Virginia governor's pronouncement and evident extent of its impact, after not more than a few moments, I found myself as a resident of the state anything but surprised. That eventual lack of surprise is an essential element of our individual and shared stories in this book.

We are first introduced to the long history of such failures of industry and government to protect the citizens of West Virginia against just these kinds of disasters in chapter 1. The "contamination experience" as it may be called, or at least the sensory indices of its near potential, are commonplace in the Kanawha Valley. This specific failure—despite important particularities—must be contextualized and understood as part of a long-standing relationship of well-established positions among government, industry, and local people. This was ultimately an expression of *continuity* in existing social relationships, not an exception. That fact, however, which exists on a general level of historical abstraction, does not dismiss the simultaneous fact that this was a real and, for many residents like me, new experience that exceeded a personal threshold for acceptance of the status quo. Even while this was a moment of continuousness with a popularly understood, though generally downplayed, past, it was also one that would emerge for those in our project, and beyond, as a fundamental disjuncture. For some, the event became a "break" in the everyday that helped propel many people to enact lasting change within their network of relationships, as they went from a position of passive acceptance to active personal resistance and activist efforts in their communities to challenge established ways of thinking about and doing things. Among others, Jim Hatfield's conversation in chapter 8 with local musician Paul Epstein highlights how the spill could become a kind of awakening, even a tipping point, toward activism.

Just as we sensed through the personal account of Lassiter and Campbell's exposure to chemicals offered in chapter 1, we will hear in coming oral histories how disasters (in their many guises) tend to be "totalizing" events—more accurately *processes*—that affect all aspects of victims' lives.[2] This is perhaps all the more so for disasters that entail toxic exposure and the contamination experience. An emphasis on process fits well with the sensibilities of oral history and the ethnographic method, wherein lived experience is understood to acquire meaning for those involved through reflection after the fact and acquired knowledge in relationship to what are complexly unfolding events.

Given how disasters, at least in some commonly occurring and widely reported forms such as floods and earthquakes, can result in extensive damage to the environment writ large and thus in the loss of material resources most basic to the conduct of everyday life, creating immediate, substantive and quantifiable needs, disasters are popularly perceived as *physical* events focused in particular places and times. Countering this prevailing, media-influenced perception, the scholarly analysis of disaster—including the manner in which powerful players in society, such as the media itself, package or frame causes and effects of disaster—makes clear the fact that disasters are far more than simply material phenomena. Rather, they are fundamentally a part of the world of *social relations*.[3] Reading through oral history in this volume, we can see how "relationship" serves as an organizing theme for thinking about how people are meaningfully and impactfully connected to each other, to such things as land or "place" (as I refer to it in this chapter and as we have termed the focus of part II), and to industry and government.

Speaking not only of the manner in which disasters may be misleadingly framed as singular, quantifiable events—as opposed to complex, multifaceted processes—but also of how they are commonly characterized as "unexpected," anthropologist Kim Fortun asserts that "there is ignorance of history and structural conditions in such articulation. . . . These words themselves are risky."[4] It is important to understand that although disasters may do most harm to those who do not generally expect them, they do not appear wholly "by accident." Rather, in a great number of cases, disasters are quite *predictable*. When seen in this way, the Freedom Industries chemical release—a leak from a badly deteriorated storage tank that appears to have worsened over time, allowing eventual escape of the chemical from a faulty secondary containment area and into the Elk River—was no accident. Further, such disasters disproportionately assault those who are variously disadvantaged or, at least, have the most harmful effects on socially marginalized individuals and groups. We will see how the impact of the water crisis is magnified for those with few resources, and the reader will be able to explore this fact at greater depth in both Gabe Schwartzman's chapter in part III and the digital storytelling website that grew out of his allied project on the West Virginia water crisis.

Like Anthony Oliver-Smith, we can assert that disasters are produced by the workings of "an ongoing social order, of this order's structure of human-environment relations, and of the larger framework of historical and structural processes . . . that have shaped these phenomena."[5] For example, in examining Haiti as both the poorest country in the Western Hemisphere and a frequent site of disasters, Chelsey Kivland heartbreakingly shares that "While 'natural' disasters are typically considered to be unintended, and the inevitable response and recovery designed, in Haiti this pattern is turned on its head. Disaster is certain and recovery accidental. In other words, while the precarious situations of most Haitians make another disaster highly probable, the paths *out* of such disasters are always perilous and unpredictable."[6]

Although I live downstream from the Elk, where the chemical release took place, and the Kanawha Rivers—both tributaries of the great Ohio River whose water I have long reluctantly drunk—we in the downstream city of Huntington, West Virginia, did not receive a call to avoid drinking our water. Here in Huntington, as in Charleston, American Water did not shut its intake valves. Though not by following an official directive to do so, we stopped drinking tap water in my household after the governor's announcement and did not begin using it again—carbon filtered as it was before—for more than a month. Assuming there would be more chemical releases in the future and drawing on resources my household has that many do not, we have since invested in a sophisticated reverse osmosis system that delivers better-tasting water as well as some peace of mind. Though we noticed no changes in water or air quality at home in the days and weeks after the upstream release, reports came in along the Ohio River—many miles from Charleston and even downriver from Huntington—of a sickening licorice odor as the chemical plume passed. The Greater Cincinnati Water Works, a municipally owned and operated utility (unlike American Water) located more than two hundred river miles downstream from Charleston, calculated the plume's progress, closed its intake valves accordingly, and, though diluted considerably by the time the chemicals reached the city, detected MCHM at its intake in the Ohio River on January 15 after six days of waiting.[7]

When Eric Lassiter and I began our conversations about the spill and the emergent project to document its effects, I stopped to consider my relationship not only to this particular crisis but to the idea of disaster, how disaster generally and the human experience of disaster specifically, have been researched and conceptualized by scholars and others. In doing so, I came to account for influence. It isn't easy to map the turns of a life or career that shape one's sense of things. That said, I shouldn't have been surprised to find myself a part of this project and deep in the literature of disaster. As an undergraduate, I pursued a degree in human ecology, a field inclined toward careful study of the complex relationships between humans and their varied environments. Environmental psychologist Rich Borden helped me to appreciate the relationship between experience, environment, and identity; his own experience of studying how people in the midwestern United States responded to the devastation of tornados proved particularly influential.

I built on these human ecological roots as an anthropology graduate student at the University of Michigan. On a Fulbright to Indonesia, I researched a long-standing program of government-sponsored resettlement that relocated people away from the more densely populated islands at the nation's center to less populated islands at its periphery. The integrationist central government's goal was to encourage "development" in those outer islands. The village of "transmigrants," as they are called, where I did my research had been resettled to another island following the cataclysmic 1963 volcanic eruption of Gunung Agung on the island of Bali. My research had not been designed to explore the experience of that disaster; rather it had been to examine how this group of several hundred households faced the challenge of trying to obtain some measure of ecological and cultural sustainability in an unfamiliar place. Even so, I became steeped in tales of this devastating disaster and the group's remarkable recovery and renewal at the individual and communal level.[8]

During my graduate training, I studied with Sharon Stephens and Roy "Skip" Rappaport—both of whom died of cancer during the time I worked with them. Stephens and Rappaport profoundly influenced my thinking about the complex tangle of issues raised in the study of disaster, although it has taken years for me to explicitly recognize how their influence led me toward disaster research. Both scholars were deeply committed to engaged anthropological inquiry and to collaborating with people both affected and threatened by environmental hazard. Their training and my own fieldwork developed an appreciation for emergent relationships between personhood and place, broadly defined—thus, my own role in this project.

In this chapter, I begin a tailored, not exhaustive, review of the social science literature of disaster. Starting here and continuing later in the transitional interlude between parts II and III of the book, my aim is to provide a theoretical context that allows readers to connect the oral histories to that literature and to what may be a somewhat generalizable experience of disaster and, in particular, toxic contamination as a particular form. Working from my own reading of the accounts of those affected and my commitment to the collaborative nature of this project, I explore essential areas in this literature to reveal how social scientists work to make sense of complexly layered experiences of disaster and specifically to situate the experience of those exposed to toxins within a larger story of how people in other times and places have responded. Our intent is to provide readers with opportunities to personally reflect on the accounts of lived experience presented in oral history chapters in light of concepts introduced in my chapters as well as against a range of different experiences of disastrous events and their aftermath from different contexts. I would like to offer insight into multidimensional contextual factors that range from the sociocultural to the economic and political that, together, shape a landscape of risk and vulnerability that forms the nature and shapes the extent of impact of disaster. I also explore how meaning is constructed in the wake of disasters both in immediate terms of personal experience and sense making as well as interpersonally through public discussions and media representations, and in this way I pick up the central thread of chapter 1.

## Situating Disaster

The experience of those who suffer in the wake of disaster is characteristically diminished or even ignored in official calculations and accounts. If we turn, for example, to the treatment of disaster by the U.S. federal government, we see—perhaps predictably—that physical impact (most notably couched in terms of property damage and loss) prevails over consideration of the lived experience of those affected. When governors of individual states request that the president make a "declaration of disaster" in order to tap into material and monetary aid dispensed through various federal agencies, they follow official, bureaucratic definitions of disaster that require evidence of "severity" and "magnitude" based on the event's physical impact on "individuals and public facilities." These impacts, ultimately, must exceed the capacity of local and state governments to respond.[9] As Ben McMahan suggests, such designations appear "bound to specific criteria rather than human experience, and may do well to regularize disaster from an administrative perspective, but they do little to help us better understand the experience of disaster."[10] One goal of this collaborative project is to expand our understandings of what the *experience* of disaster generally, and disaster involving chemically contaminated water specifically, may mean for people in their everyday lives in an often-neglected, *long-term* unfolding of individual and social effects following initial incidents—that is, those events typically circumscribed in media accounts as "the" disaster to which, at best, there may be some anniversary returns by reporters to account in limited ways for long-term impact. Official accounts of disaster too may be misleadingly simplistic and limited in scope.

   If we look now to the basic question of what kind of disaster took place—and continues to take place—in West Virginia, is there an easy answer? The quick reply is "No." To help illustrate how this may be a difficult conceptual determination, I will start by sharing how my thinking about disaster has changed during my time in our collaborative project. I became interested some years ago in using the concept of "slow-motion disaster" as a descriptive frame for analysis to interpret social, economic, and environmental crises. If "fast-acting" disasters like hurricanes appear to be like the acute trauma of blunt force injury, then the conditions invoked through applying the moniker of slow-motion disaster suggest something like a metastatic disease. As used in popular and scholarly sources alike, the characterization of slow-motion disaster appears as an apprehension of temporal and spatial qualities of events that unfold relatively slowly—on the scale of a generation or more, potentially—that can be catastrophic for both individuals and communities on an order of magnitude typically associated with such fast-acting disasters as hurricanes. A review of widespread contemporary headlines reveals how slow-motion disaster has been used to describe such disparate crises as the spread of invasive species such as the Asian carp in the Great Lakes; the advance of lava at Hawaii's Kilauea Volcano; famine in Ethiopia; sea-level rise; the worldwide rise of chronic, noncommunicable disease; and (retrospectively) the spread of toxic contamination at sites such as New York's Love Canal.

I first applied the notion in my own work to describe protracted, multidimensional crises in communities such as Saginaw, Flint, and Detroit, Michigan, undergoing prolonged processes of deindustrialization, depopulation through outmigration, and sociospatial restructuring fueled in part by racial discrimination. This was shortly before everyone began talking about the human tragedy of Flint in 2015. It was, in fact, the insidious unfolding of a slow-motion socioeconomic crisis—some decades in the making—that created conditions for the now-infamous contamination of the city's water supply with lead and other pathogenic contaminants. Choices made in the past decade by the city's leaders—including, notably, a state-appointed "emergency manager"—were crisis driven. The crisis to which these decision makers were reacting was founded on consequential choices made by Flint's earliest leaders a century ago to become a highly contingent cog in an ascendant industrial machine by building their city, quite literally, around a General Motors vehicle assembly plant. The impact of that foundational decision, and many that have followed, has been to reduce collective resiliency and increase individual vulnerability to a disaster such as the one created by switching the source of the city's water from Lake Huron to the Flint River and failing to adequately treat it to prevent corrosion of aging lead service pipes.

Comparisons between New Orleans post-Katrina (the 2005 hurricane that struck the Gulf Coast of Louisiana especially hard) and such hollowed-out regions as Flint and greater metropolitan Detroit expose the fact that disaster—whether fast or slow—is both a process and a state. As noted by the urban sociologist James Rhodes, "the term Katrina no longer simply references the storm that surged the Gulf Coast region in September 2005; it is instead imbued with the symbolic power to articulate social, political, economic, and environmental destruction and transformation more broadly."[11] Here then, places like Detroit or, potentially, much of Rust Belt Appalachia, including the region affected by the 2014 chemical release in Charleston, may be seen as a slow-motion, socioeconomic Katrina. Such places are held—at the apparent end of a presumed developmental arc, at least—to be like disasters, or at least composed of a set of social, economic, political, and environmental crises that rise together to some disastrous magnitude. That speaks to their measure alone, I suppose.

Of course, Katrina was such a disaster—something of profound human suffering and loss—in large part due to a particular history of social and economic processes and decisions that city of New Orleans shares with places like Detroit. Broadening use of the category of disaster to apply to both hurricanes and cities points to how—in the case of Detroit and regions like it—one side of a largely political divide frames use of the term as a means to describe high levels of inequality and racial and class exclusion that result from a dismantling of social provisions accompanying the rise of neoliberal capitalism. This side finds inequality and exclusion *disastrous* for people, with tangible trauma—directly in its effects on, for example, health and well-being and indirectly through increased vulnerability. On the other side of the political divide, disaster is embraced as an effective means to what is, ultimately, a desirable end—even if that end is not always intended to come at the direct cost of human suffering. As Naomi Klein suggests in *The Shock Doctrine*, disaster capitalism, as she calls it, seeks to profit from what is characterized as the "liberating" potential of disaster in its aftermath and all its forms to clear away remnants of a postwar social compact between government, business, and the people. The intent is to make use of disaster to unleash a market-oriented development agenda carried out by private business and supported by a greatly diminished local or state government that must operate under conditions of real and, arguably, manufactured crisis.[12]

Many would characterize the West Virginia water crisis—in fact, a federally declared disaster—at least a "human-made" or perhaps "technological" crisis to distinguish it from disasters deemed "natural" in causation, such as hurricanes or droughts.[13] But it remains contested in the academic literature, however, whether a meaningful distinction can be made between what is "human" and what is "natural" typologically or causally. Further, there may be real risk in applying the concept of "disaster" to what may be based in socioeconomic crises, for example, as we do for those catastrophic events that are supposed to be of natural origin in that this may "naturalize" human-caused or "anthropogenic" processes and lend to them a sense that changes are the effect of unstoppable forces. We may deny, or at least obscure, the work of human agency. Influential scholars, including Ulrich Beck—who writes of systematic ways that contemporary societies respond to hazards and insecurities created by processes of modernization itself—have suggested that what makes today's manifold crises worthy of special consideration, if not always deserving the designation of "disaster," when compared to those either of the past or seemingly rooted unambiguously in the realm of nature, is the fact that they now constitute a larger category of "*manufactured* uncertainties."[14]

Importantly, others suggest that anthropogenic change and, possibly, degradation of the environment and associated impact on humans is by no means new. As Mary Douglas and Aaron Wildavsky demonstrate, even pollution is not a wholly novel contemporary occurrence, nor is toxic exposure and chemical contamination, in particular, a unique experience of the modern, industrial age.[15] Indeed, in the context of labor, occupational risk, and the free-market, Miriam Kaprow explains that both lay and expert knowledge (a distinction to which I return later) about connections between work-related diseases and exposure to harmful materials has been widespread in the Western world at least since the Middle Ages.[16]

Many scholars insist that we recognize important differences among what we deem disaster—though on the basis of how they are *experienced* rather than their definitive cause as either natural or human-made. Beck has recognized that events entailing massive releases of pollutants into the environment, as in the infamous *Exxon Valdez* oil spill of 1989 in Alaska and the more recent Deepwater Horizon spill off the southern coast of the United States in 2010—both of which were precipitated by tragic technical and human failures—represent wholly new forms of risk.[17] This risk, in turn, is grounded in and given distinct character by the technologically enhanced complexity of our contemporary world.

Speaking in terms of how these *manufactured* uncertainties are distinct from threats of the past, Beck describes how "they are dependent on human decisions, created by society itself, immanent to society and thus externalizable, collectively imposed and thus individually unavoidable; their perceptions break with the past, break with experienced risks and institutionalized routines; they are incalculable, uncontrollable."[18] Jody Roberts and Nancy Langston suggest that the chemicals that so often define what are generally termed "technological" disasters entailing chemical release into the environment occupy a position between "natural" and "cultural" worlds.[19] While these substances are undeniably artifacts of the culturally shaped industrial societies that produce them, many have become persistent organic pollutants, ubiquitous components of what most would think of as the natural environment that cannot be avoided. This is quite distinct from what might be the situation in a delimited, "contaminated" prior industrial site, or "brownfield," awaiting mitigation. In the former, there is no possibility of differentiating between contaminated and uncontaminated space—*all* is contaminated.

Among other things, contamination by toxic chemicals has been called "inherently uncertain" for those who experience it.[20] Lassiter has shown in chapter 1 how personally new information, ongoing sensory experience, and conflicting speculation and theorizing by local and nonlocal experts and others allowed for persistent, personal uncertainty among those affected and suggested how this uncertainty may create or deepen distrust between the affected and officials who claim to act in the best interest of these citizens. In chapter 6 we hear how a long line of friends, who came to shower over many weeks at Cat Pleska's home, which draws water from outside the affected area, would find that "while the body was cleansed, the heart and mind were not, as everyone wondered when the ordeal would end." Throughout my experience in this project, a common refrain for those affected was to ask, "When will I feel that things are 'normal' again?"

As Phil Brown (known for his work with citizen-activists in Woburn, Massachusetts, following their 1970s identification of a childhood leukemia cluster in the town) and his coauthors note, there are many uncertainties related to contamination.[21] These include previous exposures to known and unknown toxins, as well as unpredictable synergistic effects among these toxic substances and compounds associated with present exposure. An extensive review of empirical literature published in a twenty-year period documenting the effects of 130 distinct disasters found that those popularly understood as technological have at least different and often more marked physical and psychological health consequences from the results documented after recognized natural disasters.[22] In the case of toxic exposure in the aforementioned oil spills, as well as incidents such as the release of industrial chemicals from a deteriorated storage tank in the West Virginia water crisis, popular perceptions of increased, uncontrollable risk accompanied by tangible toxicological and epidemiological uncertainty—as is evident in frequently conflicting official statements regarding toxicity and risk associated with the chemicals involved—are believed to contribute to chronic psychological stress.[23]

Some researchers suggest that in some instances officials' various, inconsistent, and incongruous statements and actions in the course of disasters can constitute a deliberate "labor of confusion."[24] This research suggests, among other things, a purposeful obfuscation designed to shape the experience of people living in danger. Specifically, the results of this deliberate muddying appear intended to maintain the status quo of powerful interests. In a similar manner, in his discussion of the environmental health impacts of Hurricane Katrina, Scott Frickel dismisses the notion of uncertainty in such contexts, focusing instead on how unawareness is *actively produced* in contexts such as post-Katrina New Orleans. Specifically, Frickel suggests that while "historians, philosophers, and sociologists of scientific knowledge study knowledge making; seldom do scholars study the *nonproduction* of knowledge or the creation of knowledge gaps. Yet scientific work involves the interplay of these two countervailing processes, and answers to questions concerning what kinds of scientific knowledge get made by who, where, and for what purposes hinge also on 'undone science' and the consequent institutionalization of ignorance."[25] Affected West Virginia residents of MCHM contamination of their water supply directly experienced the meaning and significance of "undone science." They witnessed how officials were at a loss as to how to gauge the seriousness of the release and set "acceptable levels" for exposure given a lack of scientific study on potential health impacts of the primary chemical and a sheer absence of information for how MCHM might react with other involved chemicals within the environment at large, substances (such as chlorine) routinely used within the water treatment process, or varying makeup of various water storage tanks, and delivery piping for residential and other uses.

Kai Erikson, a sociologist who famously documented the Buffalo Creek flood—a West Virginia coal slurry impoundment failure that in 1972 released millions of gallons of wastewater in a raging torrent that roared down a valley of mine worker homes, killing 125 people and devastating an entire community—argues that our contemporary "toxic emergencies" are *different* from the forms of hazards and qualities of risk that came before in human history.[26] This difference is given to the fact that—despite media efforts to bound them to time and place—they have "no frame." That is to say, these emergencies are both temporally and spatially unsettled. This unsettled state is akin to what Cat Pleska's guests who, while grateful for the opportunity to be cleansed physically after showering with uncontaminated water at her home, left with a continuing psychological burden of uncertainty about their health and well-being. A state of "unsettledness" is common for those who experience such disastrous events involving substances that are (at least to human senses) both imperceptible and quite possibly inescapable. As is certainly true in the West Virginia water crisis, available science does not provide enough about the toxicity of substances involved in industrial processes, such as those that entail use of MCHM, to accurately specify risk of harm from exposure. Erikson has poignantly called the distinctive nature of human-induced technological disaster a "new species of trouble."[27] Erikson argues that, as the direct product of human beliefs and behaviors, technological disasters such as those involving the release of hazardous materials may have preventable causes, so that there is "always a story to be told about them, always a moral to be drawn from them, always a share of blame to be assigned."[28] In many such events, entire communities are contaminated with toxic substances that, in their unique qualities, "scare human beings in new and special ways . . . [and that] elicit an uncanny fear."[29]

Even disasters generally taken to be natural in quality and causation, such as Hurricane Katrina, may entail localized or widespread contamination events. More recently, these ancillary events were widely seen and experienced after Hurricane Harvey ravaged the industrial heartland of southern Texas in 2017. Both hurricanes caused extensive damage to vast expanses of oil and petrochemical facilities in the Gulf of Mexico and led to numerous spills and the release of complex mixtures of known and unknown chemicals that blended with flood waters to contaminate large swaths of both inland and coastal areas with toxins, well beyond the immediate sites of those chemicals' production and storage.[30]

In his oft-cited report on the Buffalo Creek flood, Erikson noted that affected people were far more distressed than might be expected following a natural disaster.[31] Although some authorities tried to characterize the impoundment dam's collapse, which came after days of heavy rain, as a natural disaster, he and other researchers discovered that local people did not consider the flood natural at all. Neither did the survivors at Buffalo Creek turn to supernatural causation—refusing to credit the flood as an "act of God," which many considered a kind of blasphemy. Rather, they took the dam's failure to be human negligence and held the coal company responsible for their extensive suffering. In both popular sources and much of the scholarly literature, disasters are characterized as either natural and thus acts of God (a label still common in insurance vernacular) or technological and thus human-caused; the authors in this book, however, are focused on how people experience and respond to what is inarguably a *crisis* affecting over time multiple dimensions of their lives and, in particular, their very sense of self and place.

In looking for a way out of the categorical dichotomy of disaster—that disasters are either natural or human-made—Steven Picou and his coauthors suggest that the task for researchers is "to identify variables explaining impacts of a disaster event on communities and individuals, some of which may be more characteristic of natural disasters (i.e., damage to the built and modified environments) and others more characteristic of technological disasters (i.e., damage to the biophysical environment)." This approach maintains an emphasis on observable, physical impact.[32] I prefer to focus on how disasters—regardless of type—lead to social crisis situations, meaning they are products of socially constructed frameworks of belief and behavior that are mobilized in the wake of a catastrophic event.[33] Such socially constructed frameworks have their own meaningful influence on the severity of disaster's impact by helping shape the extent to which different individuals and groups will be affected. The qualities of a given framework help constitute what Rachel Morello-Frosch and her coauthors describe as particular "riskscapes" within which different people live, including the actual physical landscape from which their notion of risk is figuratively derived, while facing varying degrees of danger.[34]

## Considerations of Context

Invoking anthropologist Marshall Sahlins, Anthony Oliver-Smith notes that acute disaster events and the long-term processes they set into motion make affected communities de facto and unwitting "natural laboratories," given how what he describes as the fundamental, systemic features of society and culture are "laid bare in stark relief" for examination.[35] Further, and importantly, disasters reveal points of articulation between what we may call "the local" and a larger context of state, regional, national and global structures.[36]

Thinking about particular sociocultural qualities as factors in determining whether what we would recognize as disaster will emerge out of the combination of a particular human population and potentially damaging agents—such as severe weather events or a chemical release—is essential to understanding the concept of "vulnerability" as it applies in disaster research. Speaking to what he described as "landscapes of fear," geographer Yi-Fu Tuan has said that disasters are an uncomfortable reminder to every person of our varying degrees of vulnerability. Specifically, Tuan suggests that events of seeming chance can and do "remind us of our contingency" and the real possibility that at any time "our dear and familiar ways and life itself can be terminated."[37] This holds essentially true whether you are talking about a devastating hurricane in South Texas or closure of a sprawling General Motors car assembly plant in Flint, Michigan. As a conditioning force for belief and behavior of individuals and collective actors before, during, and after a disaster event—such as the chemical release in the West Virginia water crisis or the release of lead from pipes due to a new and corrosive water source in Flint—a society's pattern of vulnerability is a necessary *precondition* for disaster or at least for the defining experience of loss so commonly associated with disaster. One may say that a disaster becomes an unavoidable outcome only within the context of a historically produced and often distinctively geographic pattern of vulnerability "evidenced in the location, infrastructure, sociopolitical organization, production and distribution systems, and ideology of a society."[38]

While noting that there is no consensus within the social sciences about what constitutes the dynamic process of socially constructed vulnerability, Susan Cutter and her coauthors suggest we may think of this vulnerability as a "multidimensional concept that helps to identify those characteristics and experiences of communities (and individuals) that enable them to respond to and recover from environmental hazards."[39] Of course, these hazards are themselves a product of how a particular environment has been constructed over time through both deliberate human intervention and the unintended effects of our actions. As anthropologists who explore the phenomena of disaster understand, it is the prevailing *culture* that shapes how some people will be at greater risk of personal and collective loss due to disaster. Those who end up being more vulnerable are at greater risk at least in part because cultural values define socially meaningful distinctions between people. These may be made on the basis of such categories as ethnicity, class, or occupation.

Once made, these social distinctions can present very different degrees of risk to people by, for example, effectively sorting members of one social group from another within a riskscape and putting some people more directly in the path of harm. Here Steve Lerner's work in what have been called "sacrifice zones"—featuring stories of so-called accidental activists much like those born of the crisis in West Virginia—is especially resonant. Particular geographic locales, which came to be known as "sacrificed" in the post–World War II Soviet Union, were populated areas irretrievably polluted by nuclear waste and fallout produced during decades of the Cold War. People living in these zones were effectively given up as necessary casualties of that war. Lerner asks us to consider how chronically poor, marginalized populations who suffer environmentally induced ill health in particular areas of the United States are similarly ignored by government as well as the corporations responsible for their exposure to toxins.[40] Are these people and places impacted by environmental or economic devastation not also effectively sacrificed to outside environmental and economic interests and silenced by those in power residing outside of such zones? Lerner encourages us to see that disproportionate environmental suffering is no accident.

Much of the interdisciplinary literature on disaster has focused on the *economic* particulars of context. This is especially true for studies of disasters deemed technological given that contamination events are by-products of an economy heavily dependent on industrial processes that use a wide array of polluting, potentially toxic, substances. Some of the research on which this literature is based focuses on small-scale communities whose principal existence is tied to the harvest and use of renewable natural resources, such as fish or other wildlife, that are then threatened by such hazards as the discharge of large quantities of oil, as was the case in the aforementioned *Exxon Valdez* and Deepwater Horizon disasters.[41] These studies suggest that psychological stress and ill health is related not only to the actual loss of these resources but also to the *threat* of their loss due to disaster.[42] In other words, the experience of being at risk itself is enough to cause stress-related health problems.[43]

Other research of particular relevance to West Virginia—a state historically dependent on coal that remains economically, politically, and culturally tied to that industry in a host of ways—examines what has been described as the "material and symbolic entanglement" between communities and industries that shapes, in some cases, far-reaching economic as well as sociocultural dependencies.[44] As we shall hear in the oral history of Dave Mistich in chapter 8, "I think that events such as [the 2014 chemical release are] . . . living metaphors . . . that explain wider things that need to be talked about." These "wider things" involve entanglements that practically and meaningfully interconnect family, industry, and community. Studies such as the one medical anthropologist Merrill Singer conducted in a region running from Baton Rouge, Louisiana, following the Mississippi River southward along what many refer to as "Chemical Alley" or "Cancer Alley" show how the economic and political power of industrial actors—such as the many chemical companies operating facilities along this corridor—directly or indirectly intimidate local citizens and effectually force communities with few alternatives to accept the provision of jobs under nearly any conditions.[45]

Among other things, Singer's study suggests how local understandings and attitudes about how physical health may be impacted by nearby chemical production are conditioned by "prevailing cultural messages" regarding safety and quality of life from powerful business voices largely unified with local and state government leaders. Essentially, enhanced daily risk (however conceived) and the potential for disaster are accepted, though perhaps unacknowledged, as tradeoffs for access to employment and local economic development. Surely this is one reason that Charles Perrow said, plainly, in his suggestively titled book *Normal Accidents* that in disasters, "the issue is not risk but *power*."[46] And much of that power is expressed through an ability to instill particular forms of fear that reliably shape certain beliefs and behaviors. I turn now to the role of power not only as something embedded in structural relations, as in Chemical Alley, but also as constituted through everyday practices that shape and can be used to deliberately manipulate perception and, ultimately, meaning.

## Construction and Deconstruction of Meaning

Social theorist Pierre Bourdieu suggests that *symbolic* power may be thought of as "that invisible power which can be exercised only with the complicity of those who do not want to know that they are subject to it or even that they themselves exercise it."[47] The oral histories in this volume—including those of Cat Pleska and Joshua Mills in part II—suggest how the history of West Virginia has been shaped by both real and symbolic power of industrial actors in ways that tend to normalize risk in the context of everyday lives—lives that grew economically dependent on that industry. Even after years of decline in the economic impact of the coal industry through direct employment and ancillary business, many families remain firmly connected to the *cultural identity* that it continues to provide through personal and family history. This identification is something that the West Virginia Coal Association actively cultivates through its marketing group Friends of Coal, formed to construct (or reconstruct) an image of West Virginia's past and future economy and identity as inexorably bound to coal production.[48]

Along similar lines, based on ethnographic fieldwork in the fittingly named Argentine shantytown Villa Inflamable, which is subjected to persistent, high levels of environmental contamination as a result of its proximity to one of the country's largest petrochemical complexes, Javier Auyero and Débora Swistun suggest that perceptions about environment, including its potential toxicity, should be "analyzed as products of individual and collective biographies, as sedimentations of actors' previous place-based experiences. . . . Toxic beliefs, or to put it in phenomenological terms, toxic experiences, are rooted in the interactions and routines that characterize a particular place. But perceptions of hazards are also manipulable, that is, they are susceptible to being molded by the practical and discursive interventions of powerful actors."[49] As previously suggested, much of the literature on disaster attempts to address what is meant by *risk* as used in different contexts and by a variety of different actors. Social scientists are interested in how experts, laypersons, policymakers, and even communities as a whole assess hazards and construct an understanding of risk. Additionally, they explore how such varied factors as socioeconomic class, gender, age, social identity, and even religious affiliation can influence perceptions of risk. Risk may be productively thought of as unrealized harm or what Beck suggests is the "anticipation of catastrophe," where risks are "always events that are not yet real."[50]

In this sense, risk entails the *potential* of loss—in whatever manner this potential may be measured and in terms that may range from the cold, quantitative calculus of economics to expressions of existential dread by those whose welfare is personally threatened. Paul Barnes suggests that prevailing notions of risk may be distinguished as either of two essential forms.[51] On the one hand, we have reductionist approaches where the notion of risk is treated as effectively context *free* and typically expressed numerically in terms of probabilities. Such an understanding lends itself well to regulatory, public health, and actuarial frameworks. On the other hand, however, we have fundamentally experiential and thus context-*dependent* notions of risk that are based on actual people's everyday exposure to threatening situations.

Melissa Checker notes that perceptions of risk held by those who are (or who feel) threatened tend to contrast sharply with official risk assessments.[52] Anna Willow invokes this tension in her recent examination of hydraulic fracturing used to obtain shale gas deposits, including in areas throughout Appalachia. Willow suggests that ongoing ambiguity about effects of this drilling method—such as exposure to vented gases or undisclosed proprietary chemicals used in the process—allows both sides of the ongoing debate to "engage in highly politicized 'discourses of justification' that (mis)represent real risks."[53] Further, she specifically proposes that pro-fracking narratives, supported by powerful business interests frequently backed by various levels of government, position what amounts to a neoliberal economic cost-benefit assessment as the "only legitimate discourse on the subject, thus marginalizing those who interpret fracking's impacts in *non-economic* terms."

Such non-economic terms could bring in the multitude of experiences associated with living in places affected by the latest industrial boom in resource extraction, which includes growing swaths of West Virginia and surrounding states that have significant Devonian black shale deposits (known as the Marcellus Shale Formation) from which natural gas can be derived. Along these lines, consider how in his study of a suburban landfill in southeast Michigan, Josh Reno argues that residents of contaminated communities attempt to "marshal their *experiences* of place as evidence to make authoritative claims about their surroundings" in ways that are often sharply at odds with official narratives in which the concerns of laypersons are frequently assumed to be at least uninformed and quite possibly irrational.[54]

Addressing this apparent impasse, the term "anthropological shock," as employed by Beck, speaks to what could be characterized as the basic *incommensurability* between scientific ways of understanding environmental risks and the lived experience of those who must coexist with these risks.[55] Similarly, Singer suggests that the incongruity between the scientific, expert, or official and thus dominant voices wielding symbolic power on the role of industrial pollution and the origin of health problems on the one hand and the place-based experience of people in affected communities on the other may result in what Auyero and Swistun refer to as a state of "toxic uncertainty."[56] In this simultaneously cognitive and emotional state, affected people may embrace—to varying degrees—what amounts to misinformed or unsound scientific understandings; this could include victim-blaming that identifies their own or others' lifestyle choices as causally responsible for their health problems and even an outright denial or acceptance of risk.[57]

As noted, disasters can be revealing. Among other things disasters expose how, through a variety of social interactions between affected laypersons, various government officials, scientific experts, and media representatives, people construct threats and create what Lee Clarke refers to as a "risk frame."[58] Importantly, Auyero and Swistun explain that although such frames—cognitive structures or schemata that entail what people see, what they do not see, as well as what they know, do not know, misinterpret, or even ignore about the dangers around them—are socially produced, this is not a cooperative production.[59] Rather, this process may be manipulated through both material and discursive forms of power that shape the availability and nature of information about the origins and effects of toxic contamination. For example, in her exploration of how the 1986 nuclear disaster at Chernobyl in Ukraine drew into question both state bureaucracy and scientific expertise within the former Soviet Union, Krista Harper explains that this led to a "politicization of knowing," in which local knowledge based on immediate perception of events is especially devalued.[60]

I first came to appreciate the scope of the Chernobyl disaster's impacts, both on the ground and in terms of social theory, while working with the late Sharon Stephens. Stephens was an anthropologist conducting research among the nomadic Sami people in Norway, Sweden, Finland, and Russia on what might be described as their sense of dislocation and ultimately *deprivation* as a result of the contamination of their reindeer herds, which are a fundamental element of their cultural identity, by windborne radioactive fallout from the explosion at the nuclear powerplant at Chernobyl in what was then Soviet Ukraine. Essential to my own emergent understanding was the fact that the consequences of this disaster were indeterminate, unpredictable, and resulted directly from human decisions. I could see that, as Stephens described, in the context of vast technological changes within contemporary society that made both nuclear power and the Chernobyl disaster possible, "our senses are increasingly expropriated from local communities, as the reality of unseen dangers is defined, negotiated, and administered by experts in the weather services, mass media, cabinet officers, radiation commissions, and laboratories."[61]

Ulrich Beck speaks of the horror of this fundamental expropriation as a "loss of sovereignty over our senses [and] . . . a collapse of everyday knowledge" that leads to an "end of perceptiveness and the beginning of a social construction of risk realities [wherein] . . . information equals reality and thus reality can be created and transformed in the shaping of information and information policy."[62] Is it any wonder that the victims of the West Virginia water crisis could be made to feel unsure of their own perceptions and unable to return to any sense of normality? Beck describes how Chernobyl became a "media event" that only further disenfranchised victims given that radiological contamination, for the very reason it escapes sensory perception, turns everyday life into a kind of "political football" for both experts and mass media. This situation is made only more challenging by social media and our 24-7 news cycle—things Beck did not have to address when he was writing in the late 1980s.

Speaking to the thesis of "social amplification of risk" that Roger Kasperson and his coauthors put forward, which states that seemingly impersonal hazards in the world must interact with psychological, social, and cultural processes in ways that variously amplify or, conversely, attenuate how people perceive and respond to risk or disaster events, Roy Rappaport suggests that "social and cultural processing of information is intrinsic to all transmissions concerning risk."[63] Further, Rappaport asserts that there is no such thing as a pure, objective assessment "free of 'social amplification.' . . . The most objective assessments, after all, are based upon the heuristics assumed valid by a particular scientific subculture."[64] Risk estimations—of all kinds—depend on the availability and quality of information as well as reference points provided by culturally informed cognitive "maps" of a given situation.

Gregory Button has done much to illuminate how oft-unchallenged assumptions entailed specifically in media-based framing mediate how everyday people think about events and construct such maps in attempt to understand, relate to, and navigate them. Button describes frames in this context as media "packages" that present seamless and seemingly objective accounts within which a news story's focus is developed and eventually understood.[65] These consumption-ready frames serve to organize our grasp of the world and thus shape how both news producers and news consumers perceive events. In short, framing can powerfully construct and reconstruct meaning in highly selective ways that legitimize some accounts—and associated agendas—in part by diminishing or obscuring others. Similarly, Oliver-Smith and Hoffman explain that in formulating meanings for what has happened, another important aspect of this process comes to light: "Very often various interpretations of events are produced, bringing up control of definition and 'story' along with tales of praise and vilification. 'Ownership' of a disaster, that is, the right to claim that it occurred, who its victims were, and the 'true account' of events, origin, consequences, and responsibilities, often erupts as a contested form of discourse in all stages of a disaster."[66]

Importantly, these interpretations extend to the very category of risk—whether it is recognized at all and, if so, who will have the ability to set its amounts and limits and thus actively shape the contours of the particular riskscape in which people actually live and work. Further, the determination of whether, when, and where a disaster has occurred, as well as how long those affected may receive how much aid, is a function of how the disaster is represented.[67] Taking the effect of culturally informed, power-shaped understandings into account, we can see how arguments made regarding the potential for hazards from new technologies are, as Paul Barnes notes, "not just concerned with choosing a safer technology or a more stringent standard over another. They are linked to fundamental questions about the social and political meaning of technologies and their broader societal implications."[68]

In her consideration of the policy implications of particular representations of risk, Judith Bradbury describes how common use of the term *perceived risk* suggests a dismissive attitude.[69] Basically, while experts are presumed to deal with "real" conditions, laypersons' risk assessments are merely personal views—despite the fact that both entail judgment and are accordingly subject to a variety of preconceptions. It is not surprising then, as Barnes notes, that the official response to public concern over possible risk has historically been to examine the nature of what is typically taken to be "misunderstanding" or "misperception" regarding the "true" nature of the particular risk.[70]

In the case of West Virginia's water crisis, official accounts framed the public response as emotionally provoked irrationality—particularly as many refused to drink the water weeks after the chemical contamination was discovered—that led to unreasonable or unwarranted fear. This reluctance to drink the water, however—as contributors to this book point out—was based not just in fear associated with the current event but in carefully considered cumulative experience. It was both logical and rational. Residents of West Virginia's Chemical Valley who have worked in the chemical industry or have simply lived through various—and numerous—prior spills and contamination events have learned to treat official accounts with some degree of skepticism. Governmental officials, however, who work in communities like those of the Kanawha Valley, frame public response as a problem of misapprehension and "treat" that problem through a simplistic approach frequently referred to in the social science literature as "risk communication."

Reno argues that claims about the existence of noxious effects of the landfill at the center of his own ethnographic study made by waste industry officials, regulators, lawyers, health and environmental activists, and laypersons residing near the facility were *all* the product of various projects of "truth making."[71] Despite the fact that in the United States we see scientific knowledge as a one-way process where information flows from dispassionate expert to an emotional but passive public, Checker explains that those within the environmental justice community point out that those who scientifically study these events are embedded in relationships of power and subjective interests and thus are not at all as "neutral" as might be claimed.[72] While recognizing the limits of objectivity in science (and thus the accuracy of results) such activists, Checker notes, nevertheless typically maintain the need to contest scientists and experimental outcomes in their own terms by combining the local knowledge born of lived experience with tools of scientific inquiry.

## On Place

Despite a literature on willful "placelessness" in contemporary American life that perhaps reached its peak in the late 1990s—when we were told by "new economy" proponents that through the power of the internet and unleashed forces of globalization, physical location would become irrelevant to public and private life—place continues to be an essential part of the intentional construction of individual and collective identity.[73] This fact is at the fore of research by both Reno and Checker, who reinforce the centrality of place-based experience by showing how it is so, perhaps even especially so, when the relationship is less than salubrious. Anthropologist Setha Low states simply that "place is space made culturally meaningful."[74] It is also personally meaningful. The folklorist Ken Ryden provides a definition closely tied to its significance to individual persons in finding that as distinct, predictable, and culturally meaningful space, place is "an essential component of individual identity . . . [and] definition of self."[75] Linguist Barbara Johnstone notes that people's sense of self and place is "rooted in narration" of the sort found in the oral histories of this volume and that "there is a basic connection between stories and places." In human experience, Johnstone explains, places themselves "are narrative constructions and stories are suggested by places."[76] Autobiographies become intimately connected with places as individuals, regardless of particular circumstances, ultimately create, hold on to, and possibly—for reasons such as the violations of trust that come with disasters such as the West Virginia water crisis—come to reject their own landscapes.

In the oral histories and biographical details in part II, we find the coalescent theme of place and in particular changing *sense of place*. The bond that many West Virginians feel for physical and sociohistorical landscapes of the state is a near legendary source of identity and pride. Following then-governor Joe Manchin's 2005 State of the State address, "Open for Business" signs were introduced at major points of entry throughout the state. Following some popular backlash against their installation, a West Virginia University student petitioned to have the signs removed. After collecting thousands of signatures and considerable publicity, Manchin announced a call-in and internet poll to settle the matter of whether or not the signs reflected popular sentiment in the state. The outcome of polling prompted return to the earlier slogan "Wild and Wonderful," though the sometimes motto "Almost Heaven" was a close second—a reference to the 1971 John Denver tune titled "Take Me Home, Country Roads."[77]

It is no exaggeration to say that most West Virginians feel a profound connection to the place they call home. Irwin Altman and Setha Low suggest that such "place attachment," as this connection is generally referred to in the literature, "is an integrating concept comprising interrelated and inseparable aspects."[78] Basically, the concept is used to refer to people-place bonds based on thought, everyday practice, and emotion. As noted by Robert Hay, the social science literature on place attachment has developed alongside related community attachment studies that, similar to my own work with lifestyle migrants, explore decision-making processes related to how people choose places to live and how they find reasons to leave.[79] In his landmark study of the meaningfulness of place, Keith Basso informs us that our "attachments to places, like the ease with which we usually sustain them, are unthinkingly taken for granted" until, that is, something happens to reveal their profound importance.[80] With their emphasis on the significance of person-place bonds to identity, it is not surprising that studies of attachment often explore the impact of "disruptions" to these bonds as a way to better understand their varied dimensions under conditions believed to illuminate what might otherwise be an unselfconscious process or state. Such "disruptions" must certainly include effects of the West

Virginia water crisis, which forced many to rethink the nature of their relationship to place in the state. What may be dramatic discontinuities born of such events are often referred to as lived experiences of displacement and entail involuntary breaks in the attachment of person to place coupled with personal loss and challenges to sense of self.[81]

## Notes

1. See Ken Ward Jr., "Crisis 'Pulls Back the Curtain' on Water Threats," *Charleston Gazette*, January 12, 2014, accessed March 16, 2015, http://www.wvgazette.com /News/201401120054.
2. Anthony Oliver-Smith, "Anthropological Research on Hazards and Disasters," *Annual Review of Anthropology* 25, no. 1 (1996): 303–28; Anthony Oliver-Smith and Susanna M. Hoffman, "Introduction: Why Anthropologists Should Study Disasters," in *Catastrophe and Culture: The Anthropology of Disaster*, ed. Susannah M. Hoffman and Anthony Oliver-Smith (Santa Fe; Oxford: School of American Research Press, 2002).
3. Gregory V. Button, "Popular Media Reframing of Man-Made Disasters: A Cautionary Tale," in *Catastrophe and Culture*, ed. Anthony Oliver-Smith and Susanna M. Hoffman (Santa Fe, NM: School of American Research Press, 2002).
4. Kim Fortun, "Disaster: Integration," Cultural Anthropology Online, 1, accessed March 3, 2015, http://www.culanth.org/fieldsights/207-disaster-integration.
5. Oliver-Smith, "Anthropological Research," 314.
6. Chelsey Kivland, "On Accidental Recoveries and Predictable Disasters: The Haitian Earthquake Three Years Later," Cultural Anthropology Online, accessed March 3, 2015, http://www.culanth.org/fieldsights/53-on-accidental-recoveries-and -predictable-disasters-the-haitian-earthquake-three-years-later, emphasis added.
7. City of Cincinnati, "West Virginia Chemical Spill," accessed March 16, 2015, https://www.cincinnati-oh.gov/water/news/west-virginia-chemical-spill/.
8. Brian A. Hoey, "Nationalism in Indonesia: Building Imagined and Intentional Communities through Transmigration," *Ethnology* 42, no. 2 (2003): 109–26.
9. Federal Emergency Management Agency, "The Declaration Process," accessed March 14, 2015, http://www.fema.gov/declaration-process.
10. Ben McMahan, "Disaster: Provocation," Cultural Anthropology Online, accessed March 3, 2015, http://www.culanth.org/fieldsights/144-disaster-provocation.
11. James Rhodes, "Extending the 'Urban Disaster' Paradigm: From New Orleans to Detroit (and Beyond?)," in *The "Katrina Effect": On the Nature of Catastrophe*, ed. W. M. Taylor (New York: Bloomsbury Academic), 117–44.
12. Naomi Klein, *The Shock Doctrine: The Rise of Disaster Capitalism* (New York: Metropolitan Books, 2007).
13. Federal Emergency Management Agency, "President Obama Signs West Virginia Emergency Declaration," January 10, 2014, accessed March 14, 2015, https://www .fema.gov/news-release/2014/01/10/president-obama-signs-west-virginia -emergency-declaration.
14. Ulrich Beck, "World Risk Society and Manufactured Uncertainties," *IRIS* 1, no. 2 (2009): 291–99.
15. Mary Douglas and Aaron B. Wildavsky, *Risk and Culture: An Essay on the Selection of*

*Technical and Environmental Dangers* (Berkeley: University of California Press, 1982).

16. Miriam Kaprow, "Manufacturing Danger: Fear and Pollution in Industrial Society," *American Anthropologist* 87, no. 2 (1985): 342–56.

17. Ulrich Beck, "World Risk Society as Cosmopolitan Society? Ecological Questions in a Framework of Manufactured Uncertainties," *Theory, Culture and Society* 13, no. 4 (1996): 1–32; cf. Anthony Giddens, *The Consequences of Modernity* (Cambridge: Polity, 1990).

18. Beck, "World Risk," 293.

19. Jody A. Roberts and Nancy Langston, "Toxic bodies/Toxic Environments: An Interdisciplinary Forum," *Environmental History* 13 (2008): 629–35.

20. Michael R. Edelstein, *Contaminated Communities: Coping with Residential Toxic Exposure*, 2nd ed. (Boulder, CO: Westview Press, 2004).

21. Phil Brown, Steve Kroll-Smith, and Valerie J. Gunter, "Knowledge, Citizens, and Organizations: An Overview of Environments, Diseases, and Social Conflict," in *Illness and the Environment: A Reader in Contested Medicine*, ed. J. S. Kroll-Smith, Phil Brown, and Valerie J. Gunter (NYU Press, 2000).

22. Fran H. Norris, Matthew J. Friedman, Patricia J. Watson, and Christopher Byrne, "60,000 Disaster Victims Speak: Part I, An Empirical Review of the Empirical Literature, 1981–2001," *Psychiatry: Interpersonal and Biological Processes* 65, no. 3 (2002): 207–39; cf. William R. Freudenburg, "Contamination, Corrosion and the Social Order: An Overview," *Current Sociology*, 45, no. 3 (1997): 19–39.

23. Duane A. Gill et al., "The Exxon and BP Oil Spills: A Comparison of Psychosocial Impacts," *Natural Hazards* 74, no. 3 (2014): 1911–32; cf. Brown, Kroll-Smith, and Gunter, "Knowledge, Citizens."

24. Javier Auyero and Débora Swistun, "The Social Production of Toxic Uncertainty," *American Sociological Review* 73, no. 3 (2008): 360.

25. Scott Frickel, "On Missing New Orleans: Lost Knowledge and Knowledge Gaps in an Urban Hazardscape," *Environmental History* 13, no. 4 (2008): 643.

26. Kai Erikson, "A New Species of Trouble," in *Communities at Risk: Collective Responses to Technological Hazard*, ed. Stephen R. Couch and J. S. Kroll-Smith (New York: P. Lang, 1991), 17, 18.

27. Erikson.

28. Kai Erikson, *A New Species of Trouble: The Human Experience of Modern Disasters* (New York: W.W. Norton, 1994), 142.

29. Erikson, 144.

30. Francis O. Adeola and J. S. Picou, "Social Capital and the Mental Health Impacts of Hurricane Katrina: Assessing Long-Term Patterns of Psychosocial Distress," *International Journal of Mass Emergencies and Disasters* 32, no. 1 (2014): 121–56.

31. Kai Erikson, *Everything in Its Path: Destruction of Community in the Buffalo Creek Flood* (New York: Simon and Schuster, 1976).

32. J. S. Picou, Brent K. Marshall, and Duane A. Gill, "Disaster, Litigation, and the Corrosive Community," *Social Forces* 82, no. 4 (2004): 1499.

33. E. L. Quarantelli and R. R. Dynes, "Response to Social Crisis and Disaster," *Annual Review of Sociology* 3, no. 1 (1977): 23–49.

34. R. Morello-Frosch, M. Pastor, and J. Sadd, "Environmental Justice and Southern California's 'Riskscape': The Distribution of Air Toxics Exposures and Health Risks among Diverse Communities," *Urban Affairs Review* 36, no. 4 (2001): 551–78.

35. Oliver-Smith, "Anthropological Research," 304; cf. Oliver-Smith and Hoffman, "Introduction."

36. Insofar as we may either conceptually or, more practically, functionally consider disasters as surpassing predicted risks by an accepted order of magnitude, they also may be thought of as a kind of diagnostic signal that reveals underlying problems in the ability of a given society to adapt to features of either or both its natural or socially constructed environment. Disaster, as documented in any given society, may be an index of relative adaptation or maladaptation to this combined environment. Much of the anthropological contribution to the academic literature on disaster—in keeping with the long-standing comparative approach of the discipline—has considered different strategies to endure adverse conditions employed by human populations at a variety of scales ranging from small, egalitarian indigenous groups surviving what may be harsh, fluctuating conditions in marginal areas to hierarchical state-level responses to periodic calamity.

37. Yi-Fu Tuan, *Landscapes of Fear* (Minneapolis: University of Minnesota Press, 1979), 216.

38. Oliver-Smith and Hoffman, "Introduction," 3; cf. G. A. Kreps, "Disasters and the Social Order," *Sociological Theory* 3 (1985); Picou et al., "Disaster."

39. Susan L. Cutter, Bryan J. Boruff, and W. L. Shirley, "Social Vulnerability to Environmental Hazards," *Social Science Quarterly* 84, no. 2 (2003): 257.

40. Steven Lerner, *Sacrifice Zones: The Front Lines of Toxic Chemical Exposure in the United States* (Cambridge, MA: MIT Press, 2010).

41. See e.g., J. S. Picou and Duane A. Gill, "The Exxon Valdez Oil Spill and Chronic Psychological Stress," in *Proceedings of the Exxon Valdez Oil Spill Symposium: American Fisheries Symposium*, ed. F. Rice, R. Spies, D. Wolfe, and B. Wright (Bethesda, MD: American Fisheries Society Symposium, 1996).

42. C. M. Arata, J. S. Picou, G. D. Johnson, and T. S. McNally., "Coping with Technological Disaster: An Application of the Conservation of Resources Model to the Exxon Valdez Oil Spill," *Journal of Traumatic Stress* 13, no. 1 (2000): 23–39.

43. Paul H. Barnes, "Approaches to Community Safety: Risk Perception and Social Meaning," *Australian Journal of Emergency Management* 1, no. 1 (2002): 15–23.

44. Auyero and Swistun, "Social Production."

45. Merrill Singer, "Down Cancer Alley: The Lived Experience of Health and Environmental Suffering in Louisiana's Chemical Corridor," *Medical Anthropology Quarterly* 25, no. 2 (2011): 141–63.

46. Charles Perrow, *Normal Accidents: Living with High-Risk Technologies* (Princeton, NJ: Princeton University Press, 1999), 360, emphasis added.

47. Pierre Bourdieu, *Language and Symbolic Power* (Cambridge, MA: Harvard University Press, 1991), 164.

48. Shannon Elisabeth Bell and Richard York, "Community Economic Identity: The Coal Industry and Ideology Construction in West Virginia," *Rural Sociology* 75, no. 1 (2005): 111–43; cf. Brian A. Hoey, "Capitalizing on Distinctiveness: Creating West Virginia for a 'New Economy,'" *Journal of Appalachian Studies* 21, no. 1 (2015): 64–85.

49. Auyero and Swistun, "Social Production," 18; Javier Auyero and Débora Swistun, *Flammable: Environmental Suffering in an Argentine Shantytown* (New York: Oxford University Press, 2009), Singer, "Down Cancer Alley."

50. Beck, "World Risk Society," 292–93.

51. Barnes, "Approaches to Community Safety."
52. Melissa Checker, "'But I Know It's True': Environmental Risk Assessment, Justice, and Anthropology," *Human Organization* 66, no. 2 (2007): 112–24.
53. Anna J. Willow, "The New Politics of Environmental Degradation: Un/Expected Landscapes of Disempowerment and Vulnerability," *Journal of Political Ecology* 21 (2014): 249, emphasis added.
54. Joshua Reno, "Beyond Risk: Emplacement and the Production of Environmental Evidence," *American Ethnologist* 38, no. 3 (2011): 517, emphasis added. See also Gregory V. Button, *Disaster Culture: Knowledge and Uncertainty in the Wake of Human and Environmental Catastrophe* (Walnut Creek, CA: Left Coast Press, 2010); Edward Liebow, *Who Is Expert at Interpreting Environmental Hazards? A Commentary on the Disabling Effects of an Expert/Layperson Dichotomy* (Emmitsburg, MD: National Emergency Training Center, 1993).
55. Ulrich Beck, "The Anthropological Shock: Chernobyl and the Contours of the Risk Society," *Berkeley Journal of Sociology* 32 (1987): 153–65; cf. Krista M. Harper, "Chernobyl Stories and Anthropological Shock in Hungary," *Anthropological Quarterly* 74, no. 3 (2001): 121.
56. Singer, "Down Cancer Alley," 151–52; Auyero and Swistun, *Flammable.*
57. Brian A. Hoey, "Creating Healthy Community in the Postindustrial City," in *Recovery, Renewal, Reclaiming: Anthropological Research toward Healing,* ed. Lindsey King (Knoxville, TN: NewFound Press, 2015).
58. Lee Ben Clarke, *Acceptable Risk? Making Decisions in a Toxic Environment* (Berkeley: University of California Press, 1989).
59. Auyero and Swistun, *Flammable.* Cf. Deborah Lupton, *Risk* (London: Routledge, 1999).
60. Harper, "Chernobyl Stories," 119; cf. Checker, "'But I Know It's True,'" 118.
61. Sharon Stephens, "The 'Cultural Fallout' of Chernobyl Radiation in Norwegian Sami Regions: Implications for Children," in *Children and the Politics of Culture,* ed. Sharon Stephens (Princeton, NJ: Princeton University Press, 1995), 294.
62. Beck, "Anthropological Shock," 156.
63. Roger E. Kasperson, Ortwin Renn, Paul Slovic, Halina Brown, Jacque Emel, and Robert Goble, "The Social Amplification of Risk: A Conceptual Framework," *Risk Analysis* 8, no. 2 (1988): 177–87; Roy Rappaport, "Toward Postmodern Risk Analysis," *Risk Analysis* 8, no. 2 (1988): 190.
64. Rappaport, "Toward Postmodern," 190. At the time that I came to know him, Rappaport was applying his "engaged anthropology" while serving on a working group of the National Academy of Sciences, counseling the Alaskan government on oil leasing on the outer continental shelf and advising the State of Nevada regarding the proposed storage site for nuclear waste located at Yucca Mountain. See Brian Hoey and Tom Fricke, "From Sweet Potatoes to God Almighty: Roy Rappaport on Being a Hedgehog," *American Ethnologist* 34, no. 3 (2007): 581–99.
65. Button, "Popular Media Reframing," 146.
66. Oliver-Smith and Hoffman, "Introduction," 11.
67. Jonathan Benthall, *Disasters, Relief and the Media* (London: I. B. Tauris, 1993); Gregory V. Button, "'What You Don't Know Can't Hurt You': The Right to Know and the Shetland Island Oil Spill," *Human Ecology* 23, no. 2 (1995): 241–57.
68. Barnes, "Approaches to Community Safety," 18.

69. Judith A. Bradbury, "The Policy Implications of Differing Concepts of Risk," *Science, Technology and Human Values* 14, no. 4 (1989): 384.
70. Barnes, "Approaches to Community Safety," 18–19.
71. Reno, "Beyond Risk," 517.
72. Checker, "'But I Know It's True.'"
73. James Jasper, *Restless Nation: Starting Over in America* (Chicago: University of Chicago Press, 2009); William Leach, *Land of Desire: Merchants, Power, and the Rise of a New American Culture* (New York: Vintage Books, 1993).
74. Setha Low, "Cultural Conservation of Place," in *Conserving Culture: A New Discourse on Heritage*, ed. Mary Hufford (Urbana: University of Illinois Press), 66.
75. Kent Ryden, *Mapping the Invisible Landscape: Folklore, Writing, and the Sense of Place* (Iowa City: University of Iowa Press), 252.
76. Barbara Johnstone, *Stories, Community, and Place: Narratives from Middle America* (Bloomington: Indiana University Press, 1990).
77. Hoey, "Capitalizing on Distinctiveness."
78. Irwin Altman and Setha Low, *Place Attachment* (New York: Plenum Press, 1992), 4.
79. Robert Hay, "A Rooted Sense of Place in Cross-cultural Perspective," *Canadian Geographer* 42, no 2 (1998): 245–66; Brian Hoey, *Opting for Elsewhere: Lifestyle Migration in the American Middle Class* (Nashville, TN: Vanderbilt University Press, 2014).
80. Keith Basso, *Wisdom Sits in Places: Landscape and Language among the Western Apache* (Albuquerque: University of New Mexico Press, 1996), xiii.
81. Marc Fried, "Grieving for a Lost Home" in *The Urban Condition*, ed. L. J. Duhl (New York: Basic Books, 1963).

# CHAPTER 3

# Toward a Collaborative Ethnography

Luke Eric Lassiter

Ethnography is often described as a research method involving a particular kind of fieldwork that draws on experience-near participation, close observation of behavior, intensive interviews, and careful mapping of sociohistorical, economic, and political relationships within their lived contexts. But ethnography also encompasses a particular way of organizing, interpreting, and scripting that collected field-based research. The act of writing takes on a central role in this organizing, interpreting, and scripting (even if the ethnographic end-product is visual—say, in the form of a video). This key assumption underlying ethnography influences the words an ethnographer chooses, how she or he organizes those words into sentences and paragraphs, and how he or she hopes the reader will understand them—and in turn process a broader understanding of a particular social or cultural phenomenon, whatever that might be.[1]

A central task of ethnography, then, is to relay a particular "experience-based" understanding through writing, through explicitly interpretive organizations of collected materials. That written interpretation is, still more often than not, left solely to the ethnographer, who builds an interpretation on fieldwork-based evidence, of course, but also may frame her or his interpretation via her or his academic discipline, previous training and research, the reading of other scholarly works, and, perhaps most importantly, artful and sophisticated links and connections that resonate with larger discussions about the topic under study. Increasingly, however, ethnographic research, writing, and interpretation are organized as a team effort, a process in which research and writing is designed and carried out through collaborative frameworks. In general, such research teams are not uncommon in social scientific research (such as when anthropologists and political scientists might work together on a study of, say, a particular sociopolitical landscape of a particular region). But in the world of ethnography, team efforts are increasingly not limited to collectives of academic or professional "experts." These *ethnographic* teams invariably may also include—in addition to other experts or professionals—a broad range of community members at a given research site or sites, "invested others" who participate in designing the research, carrying out its fieldwork, as well as writing, interpreting, and disseminating in various modes (such as books, articles, videos, websites, or other media).[2]

This particular approach to ethnography, where collaboration is mobilized as both a research *and* writing strategy, is often called "collaborative ethnography."[3] Ideally, it *"deliberately* and *explicitly* emphasizes collaboration at every point in the ethnographic process, without veiling it—from project conceptualization, to fieldwork, and, especially through its writing process."[4] A critical aspect of this kind of ethnography, whether organized by a single ethnographer or an ethnographic team, is the involvement of the so-called research subjects or informants—study participants more appropriately called "consultants" (who may or may not be a part of the main ethnographic research or writing project itself)—in the cointerpretation of ethnographic texts as they develop. Simply put, the approach situates ethnography in a local review process. But this process is not just what qualitative researchers sometimes call a "member" or "participant" check, in which ethnographic findings are shared with consultants to be checked for accuracy. The process instead is meant to engage locally invested consultants in a method of cointerpretation or, more precisely, cotheorization on a particular topic—which may affect not only the developing text but how everyone involved in the process understands and processes a particular locally framed issue. Such conversations, as many have pointed out, also have the potential to bring others into larger streams of dialogue and action.[5]

Much has been written on this, from the wide range of possible applications to critiques and countercritiques of the assumptions behind collaborative ethnography.[6] So I won't belabor the point here.[7] Suffice it to say that collaborative ethnography presents us with many possibilities, not the least of which is the creation of a shared space where academic-based exploration of a particular topic can meet and mingle with more locally framed and articulated inquiry. Arguably, ethnography already does this as it mediates cross-cultural dialogue, making multiple and local meanings relevant to larger discussions about a particular topic. But collaborative ethnography focuses and amplifies this process by concentrating that dialogue within ethnography's research, writing, and interpretive methods.[8]

By doing so, it potentially shifts the agendas and applications of research from those often solely situated within particular academic disciplines to those situated within particular sets of shared relationships that may instead advance multiple—and perhaps a wider array of—agendas for research and application all at the same time. These may include specific academic or disciplinary positions (as in this chapter, for instance), but collaborative ethnography seeks to open up other positions, other ways for participating in interpretation and application (as in the next chapter by Trish Hatfield, for instance). These various positions may not all materialize at the same time or in the same way or even have the same bearing on a given project. As in any involved conversation, positions may come and go, be remembered or forgotten, prioritized or abandoned. And just as in real conversation, some positions might have more power or influence over others at any given time as that conversation evolves and changes. As many have pointed out, the critical challenge for collaborative ethnography is not just about finding common ground for collective work but also about finding ways to engage with real differences in position (which can include imbalances in influence and power) without eschewing those important differences.[9]

This is—like the oft-used definitions of collaborative ethnography—an expressed ideal. Indeed, more than a few ethnographers have pointed out the shortcomings, even failures, of realizing the promises of collaborative ethnography.[10] In the real world, collaborative ethnography is never a one-size-fits-all approach. Because it requires critical and close attention to local contexts (single- or multi-sited as they may be), collaborative ethnographic teams must be explicit about how specific kinds of peopled relationships have led to specific kinds of collaborations, which produce specific dialogues that in turn engender specific kinds of cointerpretation and associated counderstandings.[11]

Collaborative ethnography, then, is ultimately all about relationships. Which brings me to the specific relationships that gave rise to this particular collaborative ethnography. Though these relationships did not originally come together for the purpose of doing collaborative ethnography, a particular set of dialogues and their emergent understandings began to materialize in this direction and eventually would, at least in part, both form and *in*form the chapters of this book (as well as its accompanying website and audio documentaries, introduced below).

---

The Elk River spill and its evolving water crisis attracted national attention, given, as the CSB and others pointed out, its implications for issues like national security, deteriorating infrastructure, and public trust or distrust in basic services.[12] So as our oral history work unfolded, we invariably began to run into others also endeavoring to document the spill and its aftermath. Several weeks into our project, Beth ran into audio documentarian Laura

Harbert Allen while interviewing Rebecca Roth. Rebecca, it turned out, knew that both Beth and Laura were working on similar projects and thought the two should meet. Beth and I knew of Laura—we had heard her and some of her productions on West Virginia Public Radio, for example—but had never met her in person. No longer with public radio, Laura was by then an independent producer of various multimedia documentaries and was at that time beginning work on a series of audio documentaries about the crisis and about water. Beth and Laura immediately hit it off and talked about various ways the trajectories of our common work might come together to accomplish shared goals.

Around the same time Jim Hatfield called my attention to another researcher he had met from the University of California–Berkeley, Gabe Schwartzman. Gabe, a recent graduate, had received funding from the University of California to carry out postgraduate research on West Virginia's numerous problems with water pollution (such as the problems in Prenter mentioned by my neighbor the day after the Freedom Industries spill). Gabe had worked in West Virginia and Appalachia before, basing many of his multi-sited research interests in the region and, in this most recent iteration, had proposed his current research topic—perhaps serendipitously—months before the spill incident occurred. But by the time he had arrived in West Virginia, he immediately turned his attention to this latest water crisis. He collected a broad range of interviews, and one evening, when attending a meeting of Advocates for a Safe Water System, he met Jim and heard about our oral history project. Jim insisted that Gabe and I meet.

So, at Jim's behest, Gabe and I met for beers at a local pub. And, like Beth and Laura, we immediately hit it off. As I shared the particulars of our project, I learned from Gabe about his work to organize his research as an interactive website covering the crisis linking it to earlier, similar spill events in West Virginia. The more Gabe and I learned of one another's projects, the more it seemed apropos that we had met, that we should work together to find common ground for something we both cared deeply about. Indeed, as we knew many of the same people, and had carried out interviews that could clearly compliment and strengthen our respective projects, we agreed to share all of our interviews, logs, and transcripts to date, as well as work together in a similar vein as our respective projects moved forward.

A new energy seemed to infuse our oral history project as both Beth and I dialogued more with Laura and Gabe. So we decided to do what we often do in this kind of situation: we hosted a potluck dinner and invited Gabe, Laura, members of our research team (Cat, Marla, Jim, and Trish), everyone's significant others, and a host of other folks who had worked on these projects in some way or another. A large group gathered at our home, and some met for the first time. The gathering was an enormous success, especially as it seemed clear at the outset that we all shared common goals and that our respective projects would be much stronger together than carried out apart. Something meaningful seemed to be happening. Perhaps it was the Hatfields' magic again, but whatever the case, we all talked late into the evening, with small groups inside and out huddled together talking about stories collected, various findings, issues of trust and distrust, Freedom Industries and the spill, and, of course, water.

At some point that evening, Laura, Gabe, and I began to talk about how our work to produce an audio documentary, create an interactive website, and disseminate oral history, respectively, might, in some practical way, come together beyond the sharing of materials. We found that we had similar ideas about and commitments to creating works within collaborative frameworks like those found in collaborative ethnography. I suggested that we find some way to start integrating graduate humanities program students, as Jim had originally suggested, and perhaps organize a graduate seminar around the problem of research dissemination. We all agreed that this could be an important next step for organizing our shared work.

As the director of Marshall's graduate humanities program, I regularly endeavor to involve our students in projects that will engage them directly in the actual complications of doing publicly oriented work in the humanities.[13] So this seemed like a good opportunity for our project as well as our program and its students. Laura, Gabe, and I thus began meeting off and on to outline a collaboratively designed seminar that would engage a small group of graduate students in forging our oral history research into digital documentary, particularly focusing on audio documentary and interactive website production.

Our program often engages nonprofit leaders, community activists, and other professionals as seminar instructors or coinstructors, so Laura and Gabe enthusiastically agreed to help cofacilitate the course, with Laura leading us on how to produce audio documentary as a group; Gabe (who joined us via Skype) leading us in the problem of disseminating his wide-ranging research via his planned website; and me leading us in the requisite seminar goals of linking our particular discussions to larger issues of crisis, collaborative research, and the like.[14] (Beth—with her own obligations in the college of education—would join the seminar when she could.) Like many of our program seminars, this one would be small and intimate. Titled Charleston Water Crisis Oral History and Documentary, we designed the seminar with three others in mind, graduate students who had become interested in working on our oral history research in some way or another: Jay Thomas, an exceptionally well-read student of philosophy (who also owned Blues BBQ, where Jim, Trish, and I originally met to talk about this project); Joshua Mills, an accomplished student of archaeology, anthropology, and museum studies; and Emily Mayes, a bright and gifted student of English literature. Though the three were each committed to different scholarly interests and hailed from varied backgrounds, their distinct perspectives added a critical nuance to our planned collaboration.

Those plans came to fruition in ways we didn't expect, however. We spent much time, of course, reading and discussing literatures about disaster, collaborative research, and media production; reviewing and interpreting the collected oral histories from all three projects; and talking about the complexities of research dissemination—among many other things. Though it was not entirely clear to us at the time where our efforts might lead, we began our initial work to describe the spill event, elaborate the ongoing experience of living with a developing and ongoing water crisis, and explore the experiential tensions that emerged as a result (such as staying in West Virginia or leaving, issues of trust and distrust, action and inaction, and so forth). But as the seminar evolved, and as our work on the audio documentary in particular developed, it became increasingly clear that different ideas, hopes, and expectations for collaboration were at work.

Take, for example, the world of professional audio production, where ideas of collaboration are discussed regularly; as Laura explained, it often takes many different people with varying levels of expertise to pull off a piece for, say, public radio. But the process of integrating the interviewees, the documentary's so-called subjects, into the final product—which is often done by collaborative ethnographers—is just not something that many professional documentarians, especially those working as journalists, do, at least not conventionally. Fields such as journalism and collaborative ethnography, of course, often have different goals and thus different methods, interpretive approaches, and outcomes. So as these differences emerged in our own project, we came to see a whole new set of questions for the group at large: To what extent can we negotiate, perhaps even compromise, our commitments to our fields, to our professions, to the people we work with? What are the consequences when we do or don't compromise? How do we assert or relent expertise (say, in oral history research, in writing, website design, or documentary production)? When is the authority of expertise critical to the collaborative process? Can such tensions ultimately enrich our project as we move forward?

In sum, we ended up talking a lot about collaboration itself, its meaning, and its implications. Such discussions—about the deeper complications of doing collaborative research—are, as many scholars report, commonly had in most collaborative research contexts, especially as these contexts have become inundated with multiple assumptions about and expectations for collaboration.[15] Until relatively recently, collaborative researchers often positioned these emergent assumptions and expectations within metaphors of agreement, which predominantly emphasize finding common ground for forging shared interpretation. This strategy can, though effective, also gloss over *dis*agreement—or, for that matter, issues of inequality, negotiations of power relations, or differences in moral and ethical commitments. So with this in mind, many collaborative researchers now report how they work with and across *differences* to work through and describe more complex collaborations, which, of course, yield those different assumptions about and expectations for collaboration.[16] What this means for collaborative research projects like this, on a water crisis in West Virginia, is that open and explicit conversations about collaboration itself—and about the important differences that emerge—provide us with opportunities for creating collaborative projects that struggle to act within more dynamic and nuanced frameworks. At least that's what we decided as we forged ahead.

As we did so, we had to make some critical decisions about our respective goals and about how those goals would ultimately bring both our individual and collaborative work to fruition. We acknowledged, for example, that individual parts of our efforts—with their original research agendas and assumptions—must still stand on their own, as it were, as distinctive "slices" of work not smoothed over by the larger collaboration. This was, to be sure, an understanding we came to as we brought the three different projects (oral history, audio production, and website) together, but our ongoing conversations about collaboration seemed to clarify where our individual projects left off and the larger collaborative work picked up. With this in mind, Gabe Schwartzman offers a brief discussion of his website work in chapter 9, and in chapter 10 Laura describes the same about her work to produce both her own audio productions and those produced in collaboration with the larger group.

We very much wanted to complete a collaboratively framed project, one made stronger by our work together. One point of connection we all agreed on had to do with involving local review and interpretation. We all felt very strongly that our joint work would, first and foremost, remain responsive to our interviewees and the moral and ethical commitments made to them to share—and engage in dialogue about—any and all representations before material was published in any form. Moreover, we also wanted to enlarge our group effort to speak out and act on the spill and water crisis, so this approach seemed critical to accomplishing that goal. But by the time our one-semester seminar had come to its close, we had only begun to share parts of the developing audio production on a limited basis, and we actually never got to the point where we could start a more intensive process of cointerpretive review with the participants of the oral history interviews.

This development was perhaps fortuitous because it forced us into a more serious exchange about just what we wanted to accomplish beyond our joint audio production, the work for which had to extend beyond the seminar, with Laura at the helm. As we seemed to be in a good position to begin work on some version of collaborative ethnography, I successfully made the case that we organize a second seminar for the next semester, reengage with the rest of those involved in the original oral history collection, and begin work on a book project together. Thus, by the semester's close we had loosely planned a follow-up seminar for the next semester, which would focus us on collaborative writing and its attendant approaches to cointerpretation and cotheorization.

We began the writing work that eventually came to form the chapters of this book in a 2015 fall seminar titled Writing the Water Crisis. While this and chapter 1 introduce the historical, methodological, and theoretical context for writing about the Freedom Industries spill and contamination of the public water system, the following chapters assemble work that brought together Jay, Josh, and Emily with Cat, Trish, and Jim (Marla Griffith could not join us for this seminar) to write about the various themes that had emerged in the oral history interviews. As the seminar developed, we negotiated who would cover which topics, how they would go about framing these ideas, how they would approach their writing, and what forms their writing would take. Some of this writing, of course, progressed smoothly; some of it did not. But by the seminar's end, we had produced the beginnings of the final chapters that have since been edited and revised multiple times as they have passed through review both internally, among the book's authors and the larger collaborative group that included others like Gabe and Laura, and via interviewee participant checks, which also included new cointerpretations (as is common in collaborative ethnography) and focus-group-like small group reviews.[17] Beth served throughout as an editor of chapters and book flow.

We also engaged two others in the second seminar. One was Wesley Kuemmel, a graduate student well read and practiced in cultural studies, who provided some of the book's photographs. The other was Marshall University professor Brian Hoey, who had developed some of his own writing based in his research and a review of our collected oral histories. As it had become clear to the rest of us that we needed some fresh, outside perspectives, we hoped that Brian might help us make connections to larger issues of crisis, trust, and water, among other things. His analysis is offered in chapter 2 and an interlude, which include an anthropological take on both our oral history findings and the spill's larger sociocultural implications.

I should end here by noting that our book does not fall into what many might recognize as "collaborative ethnography" in fieldwork-based disciplines such as anthropology—mainly because the work that follows did not surface in what is perhaps a more conventional process of moving from, for example, fieldwork to writing initial descriptions, from writing drafts to fieldwork for cointerpretation and rewriting, and from cointerpretation to collaborative analysis and writing for final book production. Though we have deployed several approaches of collaborative ethnography—such as foregrounding ethical and moral commitments to consultants, carrying out team-based writing, and engaging in community review—in some ways, our work was much more multidirectional, with the boundaries circumscribing "the field" or "fields" set not just in an interplay of space and time, but primarily in a constellation of relationships. In this way, I do think our book surfaces in a stream of contemporary collaborative ethnography in its various and diverse forms. And as such, perhaps it will also surface as a meaningful response to Jim's original appeal that we find some way to assess the spill's impact on our community and, importantly, on the future of our public water system.

## Notes

1. Though the epistemologies and ontologies of ethnographic practice have changed dramatically in the past several decades (see, e.g., James D. Faubion and George E. Marcus, eds., *Fieldwork Is Not What It Used to Be: Learning Anthropology's Method in a Time of Transition* [Ithaca: Cornell University Press, 2009]); since the publication of Clifford Geertz's *Interpretation of Culture* (New York: Basic Books, 1973), these assumptions have arguably remained at the center of ethnographic research and interpretation.

2. For more on this particular perspective, see, e.g., Elizabeth Campbell and Luke Eric Lassiter, *Doing Ethnography Today* (Oxford: Wiley-Blackwell, 2015).

3. A full discussion of collaborative ethnography, its history, methods, and various types is beyond the scope of this chapter, though I do briefly point out some of its more salient issues here. For more on my particular approach that informs this work, see Luke Eric Lassiter, *The Chicago Guide to Collaborative Ethnography* (Chicago: University of Chicago Press, 2005).

4. Lassiter, 16.

5. Lassiter, 133–54. See also, e.g., Joanne Rappaport, "Beyond Participant Observation: Collaborative Ethnography as Theoretical Innovation," *Collaborative Anthropologies* 1 (2008): 1–31.

6. See, e.g., *Collaborative Anthropologies* (Lincoln: University of Nebraska Press).

7. I survey its arguments and counterarguments, in part, in Luke Eric Lassiter, "Collaborative Ethnography: Trends, Developments, and Opportunities," in *Transforming Ethnomusicology*, ed. Salwa El-Shawan, Castelo Branco, and Beverley Diamond (Oxford: Oxford University Press, forthcoming).

8. Lassiter, *Chicago Guide*, 133–54.

9. See, e.g., Les Field, "'Side by Side or Facing One Another': Writing and Collaborative Ethnography in Comparative Perspective," *Collaborative Anthropologies* 1 (2008): 32–50; Deepa S. Reddy, "Caught in Collaboration," *Collaborative Anthropologies* 1 (2008): 51–80; and Rachel Breunlin and Helen A. Regis, "Can There Be a Critical Collaborative Ethnography? Creativity and Activism in the Seventh Ward, New Orleans," *Collaborative Anthropologies* 2 (2009): 115–46.

10. Several articles in *Collaborative Anthropologies* take up this issue, reporting on collaboratively deployed projects with less than ideal results. At the same time, though, they offer important and critical lessons. See, e.g., Mark S. Dolson, "Reflections through Reflexivity: Why My Collaborative Research Project in Arctic Labrador Did Not Work," *Collaborative Anthropologies* 6 (2013): 201–36.

11. For more on this, see Campbell and Lassiter, *Doing Ethnography*, 15–29.

12. See U.S. Chemical Safety and Hazard Investigation Board, Investigation Report: Chemical Spill Contaminates Public Water Supply in Charleston, West Virginia, Report No. 2014-01-I-WV (Washington, D.C.: U.S. Chemical Safety and Hazard Investigation Board, 2016).

13. This position is elaborated on our program website, http://www.marshall.edu /graduatehumanities/. See especially "Toward an Applied and Public Humanities," accessed December 11, 2016, http://www.marshall.edu/graduatehumanities /files/2011/08/GH-Program_Vision_Statement.pdf.

14. A description of this seminar, "Charleston Water Crisis Oral History and Documentary," can be accessed on our program website at http://www.marshall .edu/graduatehumanities/previous-seminar-schedules/.

15. See, e.g., George E. Marcus, "The End(s) of Ethnography: Social/Cultural Anthropology's Signature Form of Producing Knowledge in Transition," *Cultural Anthropology* 23, no. 1 (2008): 1–14.

16. The scholarship on these collaborative research developments are surveyed, in part, in Lassiter, "Collaborative Ethnography."

17. For more on these particular strategies, see Lassiter, *Chicago Guide*, 139–54.

# CHAPTER 4

# Chemical Spill Encountered

Trish Hatfield

Approximately five days before the spill was officially announced, Paula Clendenin noticed something amiss with her cats. "They were driving me nuts," she recalled. "I had to let them in and out, in and out, which was not usual for them. I'm working and not paying that much attention."[1]

Paula was finishing up some new paintings. A native of West Virginia, she's a retired West Virginia State University art professor whose work can be found in collections in Europe and throughout the United States. She's primarily a painter and printmaker, and on this day her work was going "really great," as she describes it; then one by one her cats started throwing up all over the place. "I thought, 'Oh my God, I'm home, with three cats sick. I can't afford three vet bills. What am I going to do?' I watched them on the deck trying to drink out of flowerpots and licking the ice." Paula gave the cats more water but they wouldn't drink it.

"I'm mad and worried and frustrated . . . cleaning up a lot of cat vomit all day long. Then on Thursday, after I heard about the spill, I put them on bottled water, and it cleared up. I put two and two together: their smaller bodies couldn't handle the chemicals!"

## Background: Who I Am and Where I'm Coming From

As I heard Paula's story (collected by Cat Pleska, coauthor of chapter 6), I couldn't help but think about how Paula's cats signaled the presence of MCHM in ways reminiscent of how caged canaries, once used in coal mines, could sense a problem before humans could. Before the use of electronic detection devices, the tiny birds could provide an early warning signal: any signs of stress on the bird alerted coal miners to the presence of carbon monoxide before the odorless gas could overcome them. The birds provided a quick way to detect a problem before it was too late.[2] Unlike canaries in coal mines, though, in this case—as will become clear in this and in the following chapters—few, save Paula perhaps, were paying much attention to early warning signs, from cats or otherwise.

But before going any further, I should elaborate some about where I'm coming from and how I got involved in this project. I am a transplant from the flat, dry plains of Colorado. In 1978 I moved to West Virginia with my husband, Jim, who had taken a job at Union Carbide's central research facility in South Charleston, West Virginia. We didn't consider until later what it might mean to live in a "riskscape" chock-full of chemical plants (see Hoey's chapter 2 for more on riskscapes).

From the time we arrived, we loved West Virginia's green rural beauty and friendly people. Jim found his work at Union Carbide challenging and rewarding. I had recently graduated with a bachelor of arts degree, and we'd assembled a garage studio as my launching pad to artistic fame and fortune. But once our sons were born in 1979 and 1981, I found parenting a lot more fun and more creatively interactive than working alone in my studio.

One morning shortly after we moved into our new home, I stepped out on our porch to take in the valley view and immediately noticed a strange odor. (We'd just spent six weeks in a Charleston hotel, but I'd never noticed something like this. We figured out later it was because we were now living two miles downwind from a chemical plant.) Although I called the state hotline for reporting such things, there was no apparent follow-up, and I let the matter drop. As an expectant mother, I was concerned. I considered myself (then and now) an environmentalist, so the odor was doubly worrisome. As time passed, I muttered about the bad smell whenever it occurred, mostly in the early morning hours, but after some time, I slipped into acceptance. Besides, it didn't seem too different from my growing up in cattle country out West, where I also encountered a range of "perfumes" from nearby feedlots.

Then in 1984 Union Carbide's pesticide plant in Bhopal, India, leaked methyl isocyanate gas, "immediately killing at least 3,800 people and causing significant morbidity and premature death for many thousands more."[3] In response, a broader public conversation developed about the risks of living near chemical plants and what safety measures might be taken. Now I was paying a lot more attention to unusual smells from nearby chemical plants. To make matters worse, the Union Carbide plant, just up the Kanawha River and only two miles from our front door, also produced methyl isocyanate gas.

Jim and I, like many others, developed an escape plan in case a Bhopal-like leak happened here, too, perhaps naïvely thinking we would have time to escape. I would keep the car at least half full of gas so I could gather the kids, now five and three, and flee the valley on the ridge road behind our house. Somehow, somewhere, and sometime later, the boys and I would connect with Jim. This was before cell phones, of course, so meeting up might be challenging but probably the least of our worries if, indeed, such a disaster unfolded. With time, we (and seemingly everyone else) settled back into the typical mindset of "it can't happen here." We didn't get our news from TV, so the usual images that might keep the most destructive industrial accident of the twentieth century alive for me made the Bhopal disaster easier to keep at arm's length. Of course, Jim heard a lot about it at work, and it proved to be a severe blow to Union Carbide.

Fast-forward almost thirty years, and on January 9, 2014, I was visiting my sister near Seattle. The next day, Jim called to tell me about the developing water crisis. He had invited folks we knew to come to our house to stock up on water, take showers, and do their laundry. Our water in Saint Albans, only fourteen miles from Charleston, is taken from the Coal River and was unaffected by the spill.

I returned to West Virginia a few days later, concerned about the impact on the residents in and around Charleston and, in a strange sort of way, excited that the crisis had occurred in the capital city while the legislature was in session. Perhaps, I thought, our legislators might act on something that affected them urgently and directly. I hoped for public outcry, national media coverage, and individualized inconvenience that would shake our legislature out of its complacency, as it rarely acted on such things—which have been reported for years in places like the not-too-distant southern West Virginia coalfields. Perhaps the few legislators who cared, I thought, those who are keenly aware of the importance of protecting our waters, would finally find a majority position and pass some important, and much-needed, legislation.

Now at home, I watched Jim become increasingly agitated by how little the authorities seemed to know about the chemistry of safe water and MCHM and about basic technical issues like continuous chemical monitoring systems. Jim is now retired, Union Carbide is long gone from Charleston's Chemical Valley, and methyl isocyanate gas hasn't been produced here for many years. But this event radicalized my husband as his training and background as a chemist came to take on new meaning and urgency. He just couldn't stand by and watch this unfold independent of science and common sense. This event propelled him into a new role as a water activist and advocate, and his training as a chemical engineer took on new significance. (Jim has more to say about his activism and others' in chapters 8 and 11.)

In addition to being a sole proprietor of my own company (CharacterEthics, which helps organizations and groups act humanely and strategically), I work part time as a program assistant for the Marshall University graduate humanities program. Its director, Eric Lassiter, lived in Charleston with his wife, Beth Campbell. One afternoon during the first few days of the spill—when Eric and Beth came to our house to take showers and wash clothes—Eric shared with me his own frustration about not knowing exactly what he could or should do about this crisis. Like so many others, he felt powerless, he said. Remembering this, it wasn't long after that I told Eric what Jim was up to and that Jim wanted to meet, that he wanted to talk about what we, as a group, might be able to do.

So one afternoon Eric, Jim, and I met at Jay Thomas's restaurant, Blues BBQ, to discuss just this. And over lunch we decided we needed to do an oral history of the crisis, to capture in some way the experience of living through the spill and dealing with an ongoing water crisis.

---

In chapter 1, Eric describes the oral history process. One of the earliest and most comprehensive interviews I conducted was with Becky Park, a local self-described activist for good government who lives in the Charleston area.

## Eight Gallons Stored Away

"I wasn't angry. I was just really happy," Becky Park said, describing her distinctive response to the spill.[4] "I had a sense that positive good . . . would come out of this. . . . To have our people think more carefully about their water system . . . , I think it's important for people to be jolted awake if they're complacent. . . . So I am very happy with the fact that we have a lot more people paying attention to all [of these] issues."

I have known Becky for a long time, and when we met to talk about the spill, I fully expected her to say something unexpected. Her contrariness is authentic; she thinks very carefully about these things. But her capacity for analytical thinking is not all she brings to this topic. Since childhood she has listened to family stories of the inner workings of the local water company: her father was an engineer for West Virginia American Water, her brother-in-law worked there until he retired, and her first husband is still an employee after twenty-six years. So the family has talked about water a lot. In addition, she has two years of training in civil engineering technology. And she has worked for environmental justice, including, as she says, as "a citizen activist around the issue of mountaintop removal." She is currently a climate lobbyist, among other things. Importantly, though, she has a passion for organizing. One of the groups she organized, for example, calls itself the Little Old Ladies Who Love Our Land. It's a small group of older women who inspired her to learn more about coal issues and push state leaders and political candidates to diversify the economy.

Several months after the spill, we met at Becky's house for the interview. After brewing tea and catching up on family news, we turned to the business at hand. "All right," I started, "can you tell me when you first knew something was going on?"

"I remember it clearly as a lot of people probably do," said Becky. "I was downtown at the Department of Health and Human Resources facility in Charleston in the old Diamond Building. We were just wrapping up this training [when] peoples' cellphones started ringing and text messages came in." The "do not use" order had just been announced by the governor.

"Right away I knew that I had eight gallons of water stored in the basement for just such an emergency. There was a fellow [in the training] that started to collect change and small bills [from people in the class] because he wanted to go to the vending machine and buy several bottles of drinking water before he went home. So at least my household and his household didn't have to panic about how to even get a sip of water before going to the store."

Becky recalled that social media let them know what was going on elsewhere. "One friend was at Foodland," she remembered, "and she submitted a photo that was picked up by the international news channel Al Jazeera, showing the stripped shelves, and in the news right away we saw the people who were actually stealing water. They put water in their buggy and tried to leave the store without paying for it."

Within a few hours, Becky's husband purchased additional five-gallon jugs and Primo brand water. He also picked up ten gallons of spring water from Monroe County and Raleigh County. They ended up with three different brands of drinking water, spring water, and "that was a comfort to know we had that." As a bonus, Becky, her husband, and her daughter, who was living at home at the time, were able to collect five gallons of condensate water a day from their furnace. She learned later that this can be very acidic and unsuitable for drinking. They noticed that a lot of people threw out their empty water jugs, but they considered empty jugs to be useful to collect tap water for household uses like flushing toilets in case the water service went out completely. They marked these gallons with a red X to keep the good water separated from the bad.

"We didn't use the [tap] water for the most part," she continued, "other than you might forget and turn the water on and wash your hands or maybe rinse a toothbrush and then realize what you'd done. We waited to hear the official news. . . . I had a sense that the chemical that was in the water wouldn't hurt us to wash our hands and things like that."

Becky turned to how many people, herself included, experienced reactions to MCHM at different points of the crisis. One occurred some time after the spill itself, when West Virginia America Water advised its customers how to flush their lines. During this time, she said, "I began to wonder if there was an aspect of having the water heated that would exacerbate the effect the chemical was having on people's skin or in their breathing. We [heard] of school personnel that were working in their school kitchens who would experience adverse effects from dishwashers and things that created a lot of steam. I was really curious about that."

Becky was not alone in her curiosity. At one point, for example, Becky volunteered to do phone surveys for the Kanawha-Charleston Health Department about the crisis: "One of the first women I interviewed had one of the most succinct and well thought out answers for her summary response. She said that she felt like people still had questions about whether the hot water affected us differently and what was now in our hot water tanks, our systems that heat our water, and if there would be any residue in there."

After describing how their family coped with the earliest days of the crisis through the flushing, I returned to something Becky said earlier about all of this being a "positive experience." "I was really relieved," she recalled, "that finally the public was thinking about what was coming out of their taps. They'll tell you right now that the quality of tap water is very, very important. But a year ago I don't know that people had any awareness at all or even thought that it was that important. . . . Now, they feel that it's important on a real . . . gut . . . level." Becky recalled a friend saying to her, "Boy, you never really think about everything you use water for until you can't use the water."

Becky laughed out loud remembering her friend's comment, then compared it to her own experiences camping and living in the country: "Anybody that camps and doesn't have a spigot right there, where they camp, knows what it's like to work with just one gallon of water at a time. Anybody who has had their water compromised, either when they live on a well . . . [or when there] might be a sudden flood, and doesn't feel confident that [their] source of water is potable, [they have] thought about such things."

# "Yet What Am I Going to Do?"

"All the water out here [in West Virginia] has always amazed me," said Kenneth Mize.[5] "You cannot walk a mile without falling into a creek or a river, and yet we have water problems. What is wrong with this picture?"

Kenneth grew up in Long Beach, a suburb of Los Angeles. A veteran of the Vietnam War, he has lived in West Virginia since 1999. When he first arrived here, he started driving trucks and heavy equipment for a coal mine. He fell on hard times after having problems with his marriage and suffering, in his words, "a mental breakdown." He lives in Charleston now and still "suffers from severe depression." At times he calls on Charleston's Covenant House, a local nonprofit that helps people meet their basic needs.

On the day of our interview, Kenneth and I met at Covenant House. He began by telling me how he first heard about the spill from his priest and from his wife. "It reminded me of a severe emergency. Being without water . . . , you really take it for granted. And then, all of a sudden, you don't have it. . . . You have to rely on Covenant House or the fire department. You don't have transportation. You have to go down [there] and get water. Just taking a shower was really stressful."

Although the water smelled like licorice, he eventually decided to take a shower. "I figured if an arm and a leg didn't fall off," he said, "I'd be all right. But I've got to get in the water. I kept my eyes closed, held my breath, gritted my teeth, and jumped in."

Immediately after the spill, Covenant House distributed cases of water to him, his wife, and a host of others. "Covenant House was real helpful," Kenneth said. "They'd bring water over to make sure everything was OK. . . . But what a waste, you know? There's no reason we should have this experience. Not in America. Not in this day and age."

Water became available by taking containers down to a big water truck on Patrick Street Plaza. "The community really got together," Kenneth remembered. "We could go down to the fire department and they were handing out water, everybody was socializing and you know, you got to know everybody. Some people were carrying big stacks [of bottled water] and [saying to] people they didn't even know . . . 'Hell, [we'll] give you a ride. Throw yours in the trunk.' And everybody piled in. Or a pickup truck would come along, and everybody would [shout], 'Hey, can I have a ride?' [And they'd give them a ride.] Everybody worked together. It was beautiful."

At the time of our interview, Kenneth reported that his priest was still bringing in a case of large bottles of water every two weeks. "I give it to my cat and we drink it," said Kenneth. "But sometimes when I get up in the morning, I don't want to open a big bottle. I just get my [tap] water and make coffee. I mean, I can't live in fear." Kenneth paused. "But I'm still concerned. The back of my mind says 'I don't trust this.' Yet what am I going to do?"

## "It Was Worrying to Me on Their Behalf"

Linda Koval and I sat in a tiny meeting room in the Dunbar Public Library, elegantly decorated (with private funds) in memory of a library patron. A faint coal train whistle accompanied our conversation.[6]

Linda agreed to meet after a day spent tutoring biology and other subjects at West Virginia State University, where she works as an academic specialist for student support services. She lives on Charleston's West Side with her husband and son, who, at the time of our interview, was about to leave for his first year of college. Although none took ill, all three had some exposure to the contaminated water.

Linda first recalled hearing "a lot of buzz" while shopping at Kroger after work on January 9, and once home she checked the news and "then it got kinda scary," she said. She has a degree in biology, and has a sense of the potential "for a toxin invasion of her home and other [toxins] that had unknown consequences."

"My friend," she continued, "and her daughter [and a grandchild] were living in the Orchard Manor housing complex and had access to free water. But at a certain point, that's no longer available, and you have to get out and buy it. Both women are on a very limited income. They didn't have as easy access to transportation as I did. They had an unreliable car between them. My friend and her daughter had a skin rash from the water. I don't know if it was from taking showers or from ingesting the water before the warning. . . . The daughter has a preschooler, and for a long time she gave her son a bath in bottled water. She was so frightened of exposing him to potential side effects from bathing him in the water. That level of worry of exposure to toxins must be very stressful. It was worrying to me on their behalf. That's another thing. I resented that people I knew were put in the position of not having regular access to free water and being concerned that the use of tap water was going to have a negative effect on them."

Linda slipped into an imitative singsong voice, "*I know that there's water available*," she said with an ironic tone. "But," she continued, speaking as before, "it wasn't always easy access. They would tell you in the news where the water would be. But a lot of times people would get there and it would be all gone. They'd be waiting in line, and before they'd get to the water it would already be used up, given out to others. This was not widespread, but it happened often enough. Water is heavy, and it's a difficult process to carry enough water to do all the things you need to do. And you want to have plenty in reserve. [Hauling] is inconvenient for one thing and stressful for another. We didn't know the level of danger, or how long it would go on here. . . . My feeling was for the people living in any of the low-income housing here in Charleston."

Linda then turned to reflecting on how this affected people's attitude about water. "Even people who may not have been that concerned about environmental concerns prior to this crisis," she said, "probably got a wake-up call. For me, it goes back to the sixties and the warnings we got from Rachel Carson in her book *Silent Spring*, about how we can degrade our own health, our own safety, if we don't control the chemicals that go into our environment. It takes constant vigilance and being aware of what the possibilities might be."[7]

"Do you think there's any good to come out of this?" I asked.

"Oh, yeah. I think more awareness and more attention to this whole issue of water quality. I think it's a really good thing to take a hard look and have so much attention focused on it . . . asking questions . . . and not letting it go, which is what I hope will happen. When Rachel Carson wrote her book, she had many case studies that she was looking at about people's reaction to air pollution, toxins in the environment. . . . She was very careful to document as many details as she could about the effects of, say, pesticides on people's health. It had a huge impact on me, then.

"This [crisis] rises to that level. Looking at the chemical industry, and looking at its impact on us. Carson was right. You can't be too careful. You should spend the money you need to prevent this kind of invasion of toxic elements in our environment. So many times, if there hasn't ever been a problem, people feel like they don't need to address the safety issues. They feel it's not worth it . . . but it is worth it."

After a while I turned to asking about the interview process itself. "How does it feel," I asked, "to be asked these questions?"

"I think it's great to look back and record people's experiences so we have a record of what happened. It feels good to be part of the larger effort document. Documentation is hugely important."

———

"When we heard the water was officially declared safe to drink," Linda wrote months later when she read this chapter and reflected further on her contributions, "we kept on drinking our remaining supply of bottled water for one or two weeks. Then I asked my eighteen-year-old son whether he wanted to drink water out of the tap. He declined, saying he did not trust that the water was safe. So we continued to drink bottled water for a few months, but we did resume tap water for bathing and laundry when the 'all clear' was issued. I don't remember when we started using tap water for cooking. The whole crisis was a real education for the whole family. We no longer take potable water for granted, and I'm much less certain of the water company's ability to provide safe drinking water."

## On Doing the Interviews

As the oral history project developed and as I interviewed more people, my concern—as well as my understanding of their experience—expanded. The interviewees described and probed often in new ways that I hadn't thought about before, about the impact of the crisis on themselves and people dear to them, as well as the actions of those who they held responsible. I learned about the experience of suddenly not having safe water: fear, frustration, anger, and sadness. But in interviews like Linda's, I was also surprised to hear the voicing of hope, located along a continuum ranging from confidence in human reasoning to inspiration anchored in diverse faith traditions, that things could change.

I recorded our conversations in places where talk is often even-tempered—a bookstore, restaurant, conference room, an office, or a living room. So most interviewees spoke in a cadence of everyday conversation, speeding up when thought clarified, hesitating when exploring new territory. At one point, one of the interviewees I didn't know well spoke so slowly that I thought she was afraid to tell me what she really thought. Only later through further conversation did I realize that she was searching for the right words to describe a meaningful aspect of her experience.

Some interviewees revealed a cautious humor at the seeming lack of intelligence, information, and sympathy displayed by many in leadership positions. That laughter was often accompanied by raised eyebrows, head shaking in disbelief, or shoulders up and arms open.

"*What?*" said Kenneth Mize, with shoulders up and arms open. "Terrorists [could] go for [all of our] water . . . and here our own people do it. We don't need a terrorist attack; we'll just rely on our own utilities [and businesses] to do it."[8] A chuckle escaped me before I realized he was serious.

The opportunity to tell their personal stories seemed appealing to most of the people I interviewed. In fact, many interviewees seemed to radiate a tenuous excitement once we sat down together: they seemed happy to share their story and articulate their previously unrecorded insights. "I've talked about how [the water crisis] impacted Covenant House and the people we serve," said Ellen Allen, executive director of Covenant House. "But you're the first person that has [directly] asked me how it impacted me personally. There was a strong psychological impact, and when I was talking to you, I felt it emotionally. I think a lot of people felt that."[9]

Many of our interviewees often had questions about what we, the oral history research team, were going to do with the experiences they shared in the interviews, now recorded. We were, of course, very clear about intentions and about archiving their stories and writing up reports. But when I mentioned that we might publish their words in a book one day, a few interviewees decided they wanted to make their contributions anonymously. As conversation continued, though, and when they reviewed their words in draft form, these few changed their minds. Time had passed, and our effort seemed like a legitimate way to collect stories and seek collaborative understandings of a traumatic collective experience.

Still, though, the pressure of the interview event, for some, seemed daunting, especially as they were being asked to recount an experience that in some ways would "represent"—even if just partially—their community to outsiders. At the end of my interview with Kenneth, for example, I asked a question that I had also asked Linda, and most others with whom I talked: "How does it feel to be asked these questions?"

"A little unnerving," he said.

"Because?"

Kenneth's voice seemed quieter than before as he confessed that this was a first for him to be interviewed. "I'm sweatin' bullets," he finally said.

"Oh, from this interview?"

He chuckled and sighed, "Oh, my God. I just hope I've contributed something."[10]

## Water, Infrastructure, and Coal

Becky and I had been talking for some time about her encounter with the chemical spill when she brought up some of its historical dimensions.[11] Many of her earliest impressions on this issue came from her father, who worked as a water company engineer. Her impressions were reinforced later by her brother-in-law and then her first husband, both of whom worked for the same water company. Given her acquaintance with the company, Becky was careful to couch her perspective in her own experience. "I was glad that you asked me to record this story," she said. "I will give you my impressions, but they wouldn't stand up in a court of law." She laughed, acknowledging the likelihood of her own experience not being taken seriously. She pointed out that she was a daughter, wife, and sister-in-law, not an employee. What she knew of the water company was secondhand. Still, by any measure it was extensive secondhand knowledge.

"I remember that my dad was frustrated," she said, "because there were certain things that needed to be done to maintain the [water] system and they weren't considered very important by management."

"What year is this roughly?" I asked.

"I can't pinpoint when he was concerned about this one issue. But he was concerned with impressing management with the importance of flushing the lines. I don't think the general population understands what really is in our water. Certain things settle out as the water travels through the pipes, and this happens, you know, to every system everywhere. The remedy is to flush the pipes by opening fire hydrants. It should be done on a schedule. . . . It should be done regularly. . . . And it just flushes sediments out of the bottom of the pipes."

Becky and I decided this was sometime in the early eighties, soon after Becky's family moved to Saint Albans.

"Dad was tasked with training personnel," Becky continued, on "how to flush and how to determine the speed of the water coming out of the fire hydrant. He was trying to show them that if they would just measure how far the water shot away from the fire hydrant that that can be translated into the speed of the water because it relates to the force of the water coming out. It's some of the most basic formulas we learn in physics class. . . .

"I remember Dad explaining that to us, and how if he could just get the workers to understand the relationship, then they could just measure and calculate. I think he was finding it enough to ask them to measure it and record it and then let him calculate what the flow would be. That was my first awareness that maintenance was not a high priority for management."

Another source of frustration came when the water company moved its customer service call center out of state. "It was a matter of calling Alton, Illinois, and saying, 'there's a water leak on Lee Street, [or] there's a water leak on the West Side,' and knowing that we were speaking with someone who was really not familiar with where our streets were and what the neighborhoods were like."

Later she heard from several sources "that they [water company management] weren't taking care of routine maintenance issues because the water company was looking at a buyout from a German company." That rumor proved to be true when the American Water Company was purchased by one of Europe's largest energy companies, RWE, in 2001, but it was completely divested and again became American Water by 2009.

"So I'd had all these concerns with the water," Becky recalled. "As an adult and as a customer for West Virginia American Water, I knew that there's a possibility that there are things in our pipes that shouldn't be there, and that come in to our spigots and can be ingested. And I knew that the distribution system could be compromised [in a crisis] because we weren't paying attention to the condition of those pipes."

While Becky paused and sipped her tea, I couldn't help but think about a recent conversation with Commander Richard Smoot of the American Legion Riders Post 73 in Saint Albans. During a pancake breakfast fund raiser one Saturday morning at the Legion Hall, we learned that he was retired from the water company and had shared the same concerns as Becky. In its crisis coverage, the media reported that several factors went into the decision to keep the water flowing after the spill, but I couldn't remember pipe conditions being one.

Becky added another critical factor: the conglomerate nature of the system. "When our water company began to acquire public service districts, they had maps of their systems that needed to be incorporated into the mapping systems of the West Virginia American Water. I had the opportunity to ink their drawings onto Mylar so that they could be compatible with the map system that Dad was using at the time."

And then she added yet another concern to think about, pulling from her more recent experience as an activist in coal-related issues. "Since then," she said, "I've learned through my study of mountaintop removal mining that the blasting on those mine sites fractures either the bedrock or the actual locations where the wells are. This disturbs the way the water flows from people's private wells into their homes and creates the need for what used to be homes that were getting their water from private wells to be on some kind of public water system. This increased the number of households that were being served by the one intake in the Elk River. Instead of a lot of intakes and a lot of smaller systems, they all started to glom on to this one intake. It's been reported that that's one reason that there were three hundred thousand people affected."[12]

"Related to mountaintop removal mining?" I asked.

"Because so many private wells had been compromised," Becky said.

With this new information, it wasn't much of a stretch to connect this event to coal and to West Virginia's long history with mining. "The most amazing thing for me," continued Becky after a short while, was this "was the first time I saw Governor Tomblin give us instructions [about using water], and I remember very clearly that he said that this was not a coal incident." As Becky said this, I immediately thought of Ken Ward, the lead reporter on environmental issues for the *Charleston Gazette*, writing on his blog, "Coal Tattoo," that "one of the really unbelievable things it that there is even a debate about whether this is a coal-related story."[13]

Becky explained her amazement. "It was so surprising to me," she continued, "to hear the governor say that this was not a coal-related incident because everybody knew that it wasn't coal that was spilled, it wasn't coal slurry that was spilled or coal sludge that was spilled. And it wasn't a mine cave-in or workers being lost in a sludge pond when their bulldozer slid into it from the collapse of something that the bulldozer was sitting on. So on one level it's so obvious that it wasn't a coal accident. But on another level everybody knew it was a chemical that was only stored there as part of the coal process, the coal industry, coal. So it was a coal incident."

I nodded. Few people I knew actually believed the spill was "not related to coal," as the governor had stated.

Back to Becky. "Then the question in my mind was, why did the governor even bother to say that it wasn't a coal incident? The next day Senator [Joe] Manchin, our former governor, made the same statement. Again, why did Manchin even bother to say that?"

---

Several months later, when Becky reviewed and commented on an early draft of this chapter, she provided follow-up to her own question. "Tomblin's statement, that 'this is not a coal incident,' was sending [state] leadership a signal that those responsible for the spill would be investigated and prosecuted. Coal disasters, in my experience, are investigated under the guise of 'making sure something like this never happens again' and coal operators are not put out of business, even when workers lose their lives."[14]

## "They Closed the Hotel"

"I'm Keith Redmond. I've been a shuttle driver and bellman at the Marriott Corporation. I've been there nine years, and I have to say, this water thing was a devastation."[15]

Keith and I were sitting with a busy lunchtime crowd at the Bob Evans restaurant in South Charleston. We both ordered ice water, even though I nearly succumbed to banana bread. Thankfully, the waitress left us alone to talk. Keith and his wife, Alice, moved from Michigan to West Virginia courtesy of AT&T in 1998, and they both retired in 2003. Two years later he took the Marriott job.

Although they lived outside the affected area, Keith's emotional response was like that of other interviewees because of his daily exposure to employees and guests at the hotel. Like so many in the region, he and his wife opened their home so friends could access safe water.

"They closed the hotel," said Keith. "In the thirty years that hotel was there, they never closed it. We had to close it because we ran out of linens. We couldn't wash anything. No one could take a shower. No one wanted to leave, but they had no choice. There was no food. Guests were going to Walmart and bringing all kinds of fruit and coffee cakes for breakfast."

"*Employees* were bringing in food?" I asked, the din of dining causing me to miss a few words. I scoot the recorder closer to Keith.

"No, the *guests* were." Guests bringing in food was even more surprising. He waited for that to sink in with me and then Keith continued: "One day guests ordered food for everyone from Chick-fil-A in Beckley, and another day they ordered pizzas from Huntington. Guests were doing this because the hotel was full, and we were sold out when it happened. I remember it well. It was one of the busiest times of the year.

"We stayed open through Sunday noon. People that stay with us year-round were very upset. They just said they wouldn't use the water; they'd go somewhere to take a bath. But [management] still made everyone leave. Most went to Huntington, and the rest went home. It was a strange thing."

Keith turned to his work. "We were all out of work for five days," he said. "A couple of volunteers came in to man the phones. The phones just never stopped ringing." Keith explained that although Marriott paid everyone for one day off, and some employees used their vacation time, others were angry. "I mean," he continued, "our waitresses live from payday to payday, or day to day on tips. It was tough. One girl went out and got a job at Walmart immediately. And she's still working there now one day or two days a week to supplement her lost income. And even when the water was declared safe, they still weren't coming to eat at the restaurant."

After the water crisis was finally over, "we reopened on Tuesday," Keith said, "but no one came. On Wednesday a few guests checked in. Hesitantly. You know, we didn't have answers. We were just telling them what the news was telling us."

"What was the news at the time?" I asked.

"They were just saying that the water has been tested, and it is safe. But safe from what?" Keith laughed bitterly.

"How did it affect you personally?"

"Well, I was out of work, one. And two, I just didn't like talking about it. When I did come back to work, everybody I picked up, especially people arriving for the first time, would ask, 'Is the water safe?' and I didn't really have an answer. We were giving unlimited bottled water to the guests. For everything they wanted, they questioned. 'Where are your towels being washed?' I didn't feel the water was [safe], and I wouldn't drink it, and I didn't even want to wash my hands. I keep sanitizer in my pocket at all times."

"Was it frustrating?"

"Yes. It was very frustrating. I didn't have the answers for a lot people, and when I did tell them, they didn't want that kind of answer. They just thought we had a solution that would take care of everything right away. It made me angry, you know. Because we have a lot of people who come to our hotel to start a small business [and tell me], 'you wouldn't believe what they make me go through. I have to get more paperwork. I have to get insurance. I have to get all these things. . . . ' And it's just a small business, not like a chemical company . . . , and then we have this stuff out there that's not being inspected."

Keith briefly laughed. "It's just irritating to hear. . . . We have guests that joke, 'Do you glow?' It is interesting what people think and say."

I pointed out that both of us were drinking the tap water of the restaurant.

Keith justified his choice: "When we go out to other restaurants, I'm getting back to . . . but for some reason I still just drink bottled water only at the hotel because it happened there . . . , just because it happened there, and I'll always associate the two now."

Keith seemed to suggest that very little good came from this crisis: "This just gives so many people a reason not to come [here], or that are scared to come, and I don't want that. West Virginia is such a nice little state, and we shouldn't have to worry about the water and the air."

## "Mountaineers Are Always Free"

Paul Gilmer Jr. invited me to his business office on the West Side of Charleston for our interview.[16] I arrived in the middle of a downpour. His office is on a narrow, busy street, so I anticipated getting splashed; I secured my buttons all the way down my raincoat before exiting my car. I struggled with an umbrella, hopscotched the puddles, and hunched over my bag to keep the notebook and recorder dry from the blowing rain.

It all seemed worth it once I got inside, where I was met with a hearty hug from Paul, and for the first time in all the years I've known this guy, I realized that he has a life beyond how I know him as a devoted civic volunteer. It seemed his small office was wall to wall with papers, all dedicated to handling his client's paperwork. Paul's secretary swiveled around briefly and smiled when Paul introduces us. He placed a chair for me to sit in close beside his desk. Clearly, I've interrupted their workflow, but I feel welcome.

I began thanking Paul for carving out time for this interview.

"Yeah," he said, "the demands on my time are pretty much a sure thing any more, but I did this because you're my friend."

I told Paul, "I was thinking on my drive here this morning that the key to this project is relationships."

"Exactly," he said.

"Because people are so busy," I said, qualifying my observation a little, "and especially since we're so far away from the spill now."

Paul nodded his head. "It's been a while. I was thinking back to what exactly was going on in my life at the time. And it was an interesting departure from my normal life." Pausing as if to joke, Paul added, "which is never normal." During all the volunteering projects we've worked on together over the years, Paul always made the confusing parts of life seem amusing. So it came as no surprise to me, in the context of this interview, that he mixed lots of laughter with his thoughtful reflections.

A lifetime resident of Charleston, Paul raised six children and educated them all. "My Dad was an original West End family," he said, referring to Charleston's West Side (which is sometimes followed by "Best Side" when residents say where they are from). "My Mom was an East End family. They both went to Garnett High School. I've lived on the West Side for the last thirty years. I'm retired now from the IBM Corporation as well as Triana Energy. I'm a military veteran. I have a bachelor's degree from WVSU; my master's is from WVU in business administration. And I have my own business, Business Tax and Accounting Service, so I'm an entrepreneur as well. I'm very active in my community working with youth sports and other organizations like the Center for Civic Life."

Paul quickly turned to the spill: "I found out about the spill about eight hours before I was getting on a plane to go to Chicago. We weren't sure of the severity or how it was going to impact people. 'Don't drink the water' became 'Don't bathe in the water'; that became 'Don't even *look* at the water.'" He offered a hearty laugh.

"After I got to Chicago, I stayed in close touch with my family, who were pretty much upset with the fact that I could go into the hotel room to take a shower and they couldn't . . . , but they got over it. In Chicago, I'm looking at photos of sites that I know very well. . . . Then also my secretary and her three children didn't have water. This impacted them, which impacted my business, so yeah, it was an interesting time."

"Luckily," Paul added, "I have two sons here who made sure our households were fortified for drinking and bathing while we all watched [in our respective places] the reports on TV. No one got ill—except for anxiety. My nature is to not let things impact my mental or physical health, so I try to play down things. The Lord will take care of us. Everything will be all right soon. Just do those things they keep telling us to do. Listen to the Center for Disease Control, *when* they say something. Just follow instructions and everything will be OK."

Knowing Paul as I do, I suspected there was more he could say about this "interesting" crisis, so I asked, "Would you unpack a little more about why you thought the crisis was *interesting*?"

Paul leaned back in his chair and sighed. "In the beginning I thought it real interesting that nobody knew what this was. How do we have bruzillion gallons of something in a tank somewhere and we don't know what it is? I can't fathom that. I want to know what everything in my house is and I want a label on it and I want to know its impact, you know. Is it poison? Is it this or is it that?

"And to this day I don't think anybody knows what the real impact on people is. . . . There's been no studies, there's no research, or anything of *that* nature, so that was kind of an *aha* to me, like, Wow! It's not just inherent to us in West Virginia. It's all over. And people are very naïve; they take so much for granted that they are vulnerable. That was one situation that got me. The other was that how we as a people—I'll probably get in trouble for this—are in denial. Right now, today, as to why that situation occurred. We're so tied to a certain industry as our life-blood. Nobody connected the dots."

"You mean the coal industry?" I said.

"Yes. Nobody connected the dots that this malfeasance, this scraping by on the skin of your teeth to make money, is all a product of coal industry. OK? And not only that, but our leaders are in denial and will not ever say that this is a product of an industry that they think people are attacking. And if they think people are attacking an industry . . . the industry has actually been attacking them for centuries. I just don't get it. Just don't get it. That was amazing to me.

"And then another piece is the realization that this same chemical is used at the mine sites. And they wash the coal with this same stuff, which is going right down into the ground and right into the watershed that people who live in coal communities have been drinking for centuries. There's bound to be an impact on *Homo sapiens*, maybe negligible, but let's know that."

As Paul talked, I thought about the same point other interviewees, such as Linda and Becky, had made about science literacy in several other interviews.

"These are things that hit me as being odd or strange . . . ; [it] seems like there's always somewhere that somebody's concealing. I don't know why. I can guess why. Somebody does know and they are gaining benefit from it and don't want people to know."

"Were you feeling anger . . . curiosity . . . frustration?" I asked.

"I think a lot of frustration. I do commend our community, and even the water company, because there was good water that came from everywhere. The only way you wouldn't have water if you were shut in and nobody brought it to you. A lot of people did get water brought to them. This is one of the strengths of our Appalachian culture—responding to crisis. That was a very good thing, but it's not a new thing. We just do that."

At the same time, though, Paul reflected that "I was able to get a feel for the perceptions of those who are outside looking in, because of who I am and what I do. I think I was gone the next week, in St. Louis or somewhere. You become the butt of a lot of jokes: 'You bring any of that licorice smelling water with you?' That didn't bother me. Conversation is conversation, but what bothers me is that once again, here we are—'Mountaineers Are Always Free'—and we're always the butt of somebody's joke. I really get frustrated, because I don't see a commitment by our leadership to turning things around."

## On Moving Interviews to the Written Page

Throughout this oral history process, I trusted my inclination to believe that something good could come out of this water crisis. This conviction was, and still is, grounded in several sources. First, in my work, I practice an organizational development process called appreciative inquiry. It uses oral history methods to inquire into and amplify what works well in an organization or community for the purpose of affecting desired change.[17]

Second, my optimism is nurtured by an ongoing association with the Marshall University graduate humanities program, and, via Lassiter, its director, involvement in various partnerships and projects based in collaborative ethnography (explained in chapter 3). This experience has deepened my practice of appreciative inquiry. Together, appreciative inquiry and collaborative ethnography are two research strategies in which respect for the individual is a core consideration. They are unique in that they help create emotionally safe environments for people to unearth and value new understandings of shared experiences. For example, people tend to have a sense of what trustworthy leadership should feel like and what reliably clean water should taste like. Yet these ideas may never see the light of day unless they can be talked about without fear of ridicule or repercussion.

Because this oral history project surfaced in frameworks like collaborative ethnography and appreciative inquiry, one of our main concerns as a research team, as a group of friends who were often interviewing old friends or new friends for the first time, was that we stay true to the trust and reciprocity that established the project in the first place. We're not outside researchers. We live here. We want to keep our friends, old and new. We have a responsibility to them to represent their words honestly and faithfully but in ways that are not a surprise to them, in ways that do not catch them off guard, affect their relationships negatively, or cause ridicule or unanticipated repercussions.

Sometimes you can't avoid this. Sometimes you can't control the life of written words in book or other forms, no matter how hard you try. But we can hope to do the very best we can. I kept this in mind as I put these interviews to the written page and then as I shared what we had written with our interviewees. This, though, was not a straightforward process. Take as an example my exchange with Paul when we discussed his review of an earlier chapter draft. He had concerns about his characterization of the coal industry's "attacking them for centuries." Our emails back and forth went something like this:

Paul: Well now you found some I didn't remember. lol. Let's not use the one that references the coal industry.

Trish: It's such a good one, though! : ) It's so true! I'll take it out first thing. You mean this one, correct? "We, as a people, and I'll probably get in trouble for saying this, are in denial as to why that situation occurred," said Paul Gilmer Jr., who refers to the coal industry as our "life-blood." Gilmer adds, "Our highest leaders are in denial and think people are attacking this industry. But this industry has been attacking them for centuries." [The writeup is slightly different in the email because that's how it appeared in an initial draft that I shared with him a month or so early.]

Paul:   Oh OK, let it stand. I really don't have anything to lose. lol

Trish:  Are you sure? Don't want you feeling I forced you into it!! Freedom of choice here. I just stated my preference since it's a good strong statement. . . . Maybe you have something to gain. I can only negotiate with you because we're buddies and I trust that you will do what you feel is right to do. . . . You can certainly take your time. I don't need a definite until July 24.

Paul:   OK, but I have made up my mind, I said it, and I mean it, so let it stand.

Trish:  Dy-no-mite!!!! We stand together. I sent you the permission form on Tues. so you should be getting it soon. Many thanks for taking time out to do this. I know you are very busy.

Paul:   Mailed permission slip back this weekend. Glad I could help![18]

Many of us in this project had exchanges like this. Sometimes our interviewees agreed to keep quotes as they stood. Others wanted slight tweaks. Others wanted to make clarifications. Still others wanted material taken out. This is how collaborative ethnography works; it's how editing works, too.

I've made countless changes to my own words and to those of others as we've negotiated and developed the book. Yet as I move through the editing process I wonder: Is there no end to this, other than calling it quits, putting a period at the end of the last sentence written and stopping right at that moment? I feel a heightened sense of responsibility to each interviewee, and each time I listen to an interview to check its accuracy, there are new threads of the story I feel I should include.

But how to bring about closure? While trying to meet a chapter deadline, I emailed Eric about how to finish this chapter. "Time," he wrote back, "is often a key motivator, but in general, it's understood that (1) you'll keep running into new stuff each time you listen or go over materials again; and (2) you can never include everything. The key, then, becomes prioritizing, and (1) deciding what information is most important, and what should be included in the first place; and (2) when happening upon new information, asking if it rises above those priorities already chosen and included: if so, include it; if not, don't include it (it can always find its way into another project). . . . The other thing to keep in mind, too, is that the inclusion of more material can also obscure the original material by making things more confusing with too many details."[19]

I got this. I was also aware that every time I went back to editing, I started at the beginning instead of picking up where I'd left off. Not the best strategy for time efficiency, but it was like gathering my granddaughter's thick hair into one long hank, gently brushing over and over again from roots to tips to line up those pesky wayward strands that took off in all directions. I realized that I connected with the interviews more deeply this way, combing through passages, finding peace (and frustration at my own slowness in figuring out transitions to make) and order (and chaos, particularly when I realized I had duplicated one large interview unintentionally).

So the process of continual reviewing from top to bottom helped me select what seemed most important to the task, knowing each new brushing would bring a more satisfying alignment for me at the time. I don't think there is an end to learning from this, especially as I shared the written interviews with our interviewees: thoughts continued to emerge not only from me, the interviewer, but also from the interviewee, either regarding details left out or told incompletely or regarding fluctuating interpretations of experiences.

This project, then, might be best seen as a collective, collaboratively written biography or memoir. Like all memoirs, choices made about content may still, in the end, not be well received by all readers. Even with all the intentionality of reviewing material, reviewing quotes, and other word choices, there still might be some risk of anger from a family member or friend or questions from the community involved. But we've all—interviewers and interviewees—tried to, above all, remain true to ourselves.

## "Nonprofits Give Me Hope"

To many people served by Covenant House, the water crisis "was almost a nonevent because they were thinking, 'Well, so what . . . I [still] have to drink the water.' When some basic needs are not met on a daily basis anyway, there's a false choice for people," explained Ellen Allen, executive director of Covenant House.[20]

Located within two miles of both Freedom Industries and the West Virginia Water Company, Covenant House looks like a large, well-loved family home from when the neighborhood was more heavily populated with residences back in the fifties and earlier. Its soft yellow and green exterior is filled with windows, a third story nestles under its eaves, and welcoming red double doors communicate the caring activity inside. Covenant House is a nonprofit whose mission is to help the most marginalized populations meet their basic needs: food (including access to clean water), clothes, and shelter. Its staff and board practice a "fearless advocacy" for social justice on behalf of those most in need.

I'm sitting with Ellen in the same meeting room in which I interviewed Kenneth. Ellen and I are colleagues and friends, and I consider her a role model for business and social justice advocacy. She was an entrepreneur for many years before shifting to the nonprofit arena. She describes herself as primarily an advocate, who only happens to also hold the title of executive director.

Ellen had a physical reaction to the MCHM-laced tap water: it began with her usual "guzzling" of water after work on January 9, just before she learned of the "do not use" ban. But the more painful and long-lasting suffering, she said, had more to do with her mental health. "I felt like I spiraled down psychologically to . . . seeing our [leaders'] response to it. . . . I was infuriated. I felt like it wasn't a place I wanted to remain. It cut very deeply. I had no idea I would have such a visceral response. It felt like an assault. It felt like a personal assault, you know."

"What does that feel like?" I asked. "I mean, I've been assaulted, and I'm guessing that you have as well, so what does it feel like when we get assaulted?"

Ellen didn't hesitate. "One, trust was betrayed. And two, I didn't have control. I'm used to having control over many things, and that's something I didn't have control over. It still concerns me that the leadership feels comfortable to allow it to happen again."

"Do you remember anything particular that was said or just the general . . ."

"Just the general response of our leadership and the water company. . . . There seemed to be a lack of concern for how it impacted the citizens. I feel like we have resources and wherewithal to make changes and I thought about the people who have no resources."

Ellen confided that she hadn't spoken to many people about her feelings: "I found myself crying because it just seemed I'd have to leave a state I grew up in. I felt some compassion for people who've been living in the southern part of the state. This is what they've lived with for generations. I heard people talking and heard it on TV, but it just didn't impact me, so I never took it in."

Ellen's sadness also came from "this big realization that, wow, we're all very vulnerable. Here we are in the state's largest city, the capital, right here, and people in positions that can make changes are saying, 'Drink it if you want to, I don't care.'"

"Intellectually I'm [now] prepared to make a move . . . if it happens again. I'll go, I think . . . to where the water is safe." Ellen quickly noted, however, that in the same time frame as the West Virginia water crisis, Asheville, North Carolina, suffered a toxic spill when eighty-two thousand tons of coal ash spewed into the Dan River. Asheville was one of the places that Ellen and her partner, Sue, were considering as a safe place to relocate. "That was another realization. . . . It doesn't really matter where we go. . . . I guess I just had my head in the sand."

"What's some good that came out of this?" I asked.

"I think the body is very resilient. And the mind is resilient." Ellen had really spiraled into the negative, which wasn't like her. "My partner reminded me of our friendships, our advocacy, the things that are important and good about this community. I think this is probably what pulled me back.

"And I've been encouraged by the citizens organizing around this. I was emboldened by the strength and courage of the Covenant House Board by what they did in suing state entities [see chapter 8, which expounds on this suit]. But seeing what the citizens did . . . seeing what Maya Nye and People Concerned About Chemical Safety did, what Angie Rosser and West Virginia Rivers Coalition, what Margaret Pomponio and WVFREE [West Virginia Focus: Reproductive Education and Equality] are organizing. . . . These are very, very encouraging. These nonprofits give me hope that things won't be quite as easy to happen again. What nonprofits have brought to me is the faith in good people to coalesce around something they feel is good in the community. And again . . . things happen, but it's the response. Perhaps a swifter and more responsible response in the future."

"What would that response look like?"

Ellen role-plays the governor making this announcement, "'Hey, don't drink the water. We're holding WV Water and Freedom Industries accountable.'"

## Learning from Disaster

Near the end of Becky's interview, I asked her, "What water are you drinking now?"[21]

"I make coffee with tap water," she said, "and when I want a drink of water in a glass, I go to our dispenser that has the five-gallon jugs. I see [in the media] where more people are not drinking tap water."

"Perhaps because of the availability now of . . . ," I said, stopping short. I was thinking about the various sources of bottled water that have become more available in Charleston and surrounding areas since the spill. Before I finished the rest of my question, Becky picked up on my thought.

"Right, well, you know it's part of the market forces that's making that more available, and people are subscribing to the service more. I think it's amazing that our water is probably at the same level that it was prior to the emergency, but people are thinking differently about their water."

Becky recounted the story I told her years ago about how lucky West Virginians were to have as much water as they did. I grew up in Colorado and remember water shortages as a fact of life. Watering the yard outside was often limited to every other day. My dad would set out a sign in the front yard when the sprinklers were on, confirming it was our day to water so we wouldn't get a citation. Remembering that story, Becky reported how her sister in Arizona doesn't drink their tap water because of its odor. "So I understood then that not every system delivers water that people find desirable to drink. And yet we've taken that for granted here, both because of the quantity issues, and because our tap water has always won awards for being delicious, for tasting right to people."

"Yeah, and it's a good color," I said, nodding.

"Right, and I think that it is part of my concept of household and community strength for people to have an awareness of the stuff that we have so easily—the electricity, the natural gas, the water, the sewage [treatment].

"Now," she then stressed, "if you want to talk about a disaster, then imagine what would've happened on January 9 if we had been told that we couldn't flush our toilets? That's when people would've died." She paused to let me take in that scenario.

After talking a bit about the spread of disease in such conditions, Becky's tone became more reflective, as she thought out loud about lessons learned from this event. "I think that what we've gained from all of this," Becky said, "is that unfinished, incomplete plans for just such an emergency are now going to be completed. [And they will] become familiar [to] a larger number of players instead of just a binder on somebody's shelf somewhere. [These] issues belong to government, and more and more what we see is that even though we live in a democracy that's supposed to be government 'for the people, by the people, and of the people,' we sort of professionalize this in our minds and somebody else is supposed to take care of it. I'm excited about the people being individually prepared in their households. And part of that preparation is knowledge. It's understanding what the pH of water means. What different contaminants mean for how you should purify your water. Boiling is good for killing pathogens that are biological, that are alive, but it doesn't do anything for a chemical contaminant like what we were faced with. So people need to have knowledge.

They need to have supplies. They need to have phone numbers. They need to know how to interact with a government. The government then needs to be prepared with its own management plans about what to do for different emergencies." At the same time, Becky acknowledged that the "government can't be prepared for every [kind of] emergency. . . . It's impossible for the government to be prepared for everything and to keep every single person safe."

When she finished this thought, we sat in silence for a moment. Then Becky continued, "I'm not a person to panic. I'm not a person to be angry about somebody else taking care of me. I did think it was upsetting that people would appear on the news and make statements about aspects that they weren't really clear on. At one point the governor said that the water had been cleared, that we could use our water. Nobody wanted to say that the water was safe, and he said it would be our decision how we would use our water. I think people were angry that he left it to individuals and individual households to make the decision, but it was the most blatantly obviously true statement that anybody made. It is always our decision."

I interrupted Becky at that point, remembering that Governor Tomblin's comment irritated me when he said it. "What I immediately thought of," I said, "was *if* you have access. *If* you don't have a car. *If* you're not near a grocery store. *If* you're not near a distribution point then . . . it's not a choice!"

Our conversation is getting very involved now, even passionate. Becky continued: "When we faced the possible disruption of services at Y2K, at the change of the calendar from 1999 to 2000, apparently the government wanted to be prepared, and to be responsible for no disruptions. And they kept saying, 'Don't panic.' They kept telling people that we should not panic. I would get so upset by that message, for them to tell us 'not to panic.' I feel like it's their responsibility to give us all of the information that they have, and [then] it's up to the individuals to decide whether or not to panic. So to say 'don't panic' is not giving me any information. It's telling me how I should behave, and that's not their business."

Becky and I went back and forth for a bit on whether we were talking about the same statement from Tomblin and when it happened, but she believed it was possibly within a week of the disaster, when people began to question the toxicity of MCHM. "They had this rule of thumb that said it should be below one part per million," Becky reminded me. "And then another agency said they thought one part per billion would be the threshold safe level. So at that point the governor felt like he couldn't say the water is safe for babies and pregnant women to drink because they're sort of at one end of the care spectrum. All he could say was it was up for individuals to choose."

I realized that Becky and I had, at this point, been talking for nearly two and a half hours. But neither of us wanted to quit. I remembered out loud that Becky needed to get to work, and I needed to get to another interview. We both agreed that it had been a unique opportunity to have this time together to focus on this important issue. I asked if she wanted to offer any final thoughts before we ended. Indeed she did.

Her tone was reflective again, focused on the big picture, not just the water crisis. "I've been watching certain parts of the way we behave as a community," she said, "mostly because I'm so upset with how we are not valuing our Appalachian hardwood forests and our ridgetops, how we are destroying them with mountaintop removal, how our state government allows that to happen, permits that to happen, and will not listen to the communities and the individuals who are being pretty much destroyed by that activity. [But ultimately] this is a positive thing that no one was killed during that disaster with the water. We have seen so many more people become publicly engaged with these issues that it gives me a huge amount of hope. I've especially enjoyed my conversations with my friend and your husband, Jim Hatfield, about the need to not just place blame but to establish what the community wants in terms of a good system."

Becky had turned her thoughts back to the water system in particular. She reminded me that "we have to make decisions in all these things about where we would invest our limited dollars, and so when we say, 'Well, we think the water system should be completely prepared for disasters,' we have to define which disasters, and how much we're willing to spend. The only way that I can have hope about our society these days is [to consider] that our culture is somewhere on a continuum of what its lifetime is, [what] its maturation process [is]. . . . If we would think of our culture here in West Virginia and in the United States in general as a person, do you think of our culture as being infantile? Adolescent? In its teenage development years? Pre-adult, young adult, adult, middle-aged, maturing, old age, maybe senility?"

"Hmm." Her question caught me off guard.

Becky continued: "I think the way we run through our resources without thinking [seems to suggest that] we are at a mid-teenage [development], I'd say, like starting to become adult, starting to make an assessment of how we're going to establish ourselves as a sustainable culture. When you look at the cultures in Europe, by and large you see nations who admit that they understand what their resources are. And they've learned over hundreds of years to exist within those limitations. Our energy needs have kind of thrown a wrench into [their] sustainability . . . , but I think when we look to those kind of countries, we see more mature societies that have already grappled with [deciding] how many people should be in [their] geographical limitations and how [they're] going to deal with what limited resources [they] have. I think here in the United States and in West Virginia we're *just* starting to be aware that we are limited and that we need to make public decisions about what we're going to do about those limitations. So . . . I have hope that we're going to mature."

We smiled and looked at each other for a long moment.

## Notes

1. Paula Clendenin, interview with Cat Pleska, July 8, 2014.
2. See, e.g., "A Pictorial Walk through the 20th Century: Canaries, *United States Department of Labor Mine Safety and Health Administration,* accessed July 2, 2017, https://arlweb.msha.gov/century/canary/canary.asp. For information on how the birds were cared for and valued by the miners, see, e.g., Kat Eschner, "The Story of the Real Canary in a Coal Mine," Smart News, Smithsonian.com, accessed August 17, 2017, http://www.smithsonianmag.com/smart-news/story-real -canary-coal-mine-180961570/.
3. See, e.g., Edward Broughton, "The Bhopal Disaster and Its Aftermath: A Review," Environ Health, May 10, 2005, accessed August 2, 2017, https://www.ncbi.nlm .nih.gov/pmc/articles/PMC1142333/.
4. Becky Park, interview with Trish Hatfield, September 10, 2014.
5. Kenneth Mize, interview with Trish Hatfield, September 10, 2014.
6. Linda Koval, interview with Trish Hatfield, August 12, 2014; with revisions, personal communication with Trish Hatfield, February 2, 2018.
7. Rachel Carson, *Silent Spring* (New York: Houghton Mifflin, 1962).
8. Mize, interview with Trish Hatfield, September 10, 2014.
9. Ellen Allen, interview with Trish Hatfield, September 10, 2014.
10. Mize, interview with Trish Hatfield, September 10, 2014.
11. Park, interview with Trish Hatfield, September 10, 2014; with revisions, personal communication with Trish Hatfield, July 30, 2017.
12. In a follow-up interview, Becky offered a clarification on this point: "The main issue is, because of mining practices, many wells were compromised, and public water was brought in to serve households who previously had private wells. Imagine going from terribly polluted well water to treated water from a public system, and then finding out that, too, was contaminated. It would be interesting to know how many households had to stop using their own well water, were 'rescued' by PSDs [public service districts], and then were absorbed by WVAm Water. And it's not just the blasting, it's the way they dispose of industrial waste in the coalfields. It's not just cracks in the bedrock, it's the sludge and even now fracking wastewater that is migrating into private wells" (Park, personal communication with Trish Hatfield, September 1, 2017).
13. Ken Ward, "Coal Tattoo" (blog), accessed September 22, 2017, http://blogs .wvgazettemail.com/coaltattoo/about/.
14. Park, personal communication with Trish Hatfield, July 30, 2017.
15. Keith Redmond, interview with Trish Hatfield, July 14, 2014.

16. Paul Gilmer Jr., interview with Trish Hatfield, September 11, 2014.

17. See, e.g., AI Commons: Introduction to Appreciative Inquiry, accessed August 2, 2017, https://appreciativeinquiry.champlain.edu/learn/appreciative-inquiry -introduction/. Appreciative inquiry is as much a spirit of appreciation and curiosity as it is a process. It cultivates a disposition to ask questions that help people make explicit the attitudes and conditions they want more of (in this case, trustworthy leadership and reliably clean water). The outcomes draw on the strengths already existing within individuals, organizations, and communities.

18. Gilmer, personal communication with Trish Hatfield, August 12–17, 2017. References to personal communications that follow original interviews in this and other chapters denote changes made to original quoted material.

19. Eric Lassiter, personal communication with Trish Hatfield, August 24, 2017.

20. Allen, interview with Trish Hatfield, September 10, 2014.

21. Park, interview with Trish Hatfield, September 10, 2014.

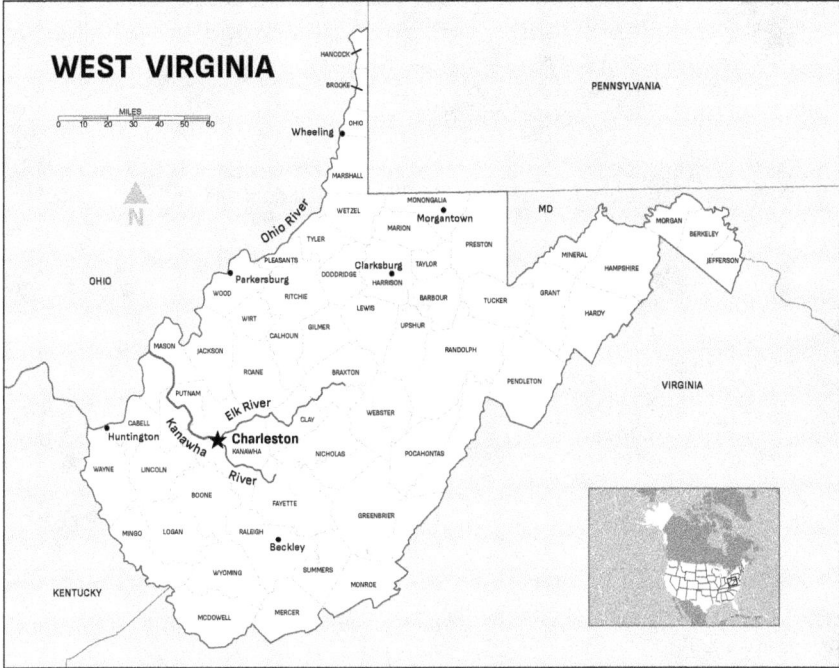

West Virginia and surrounding states. (Map by Than Saffel.)

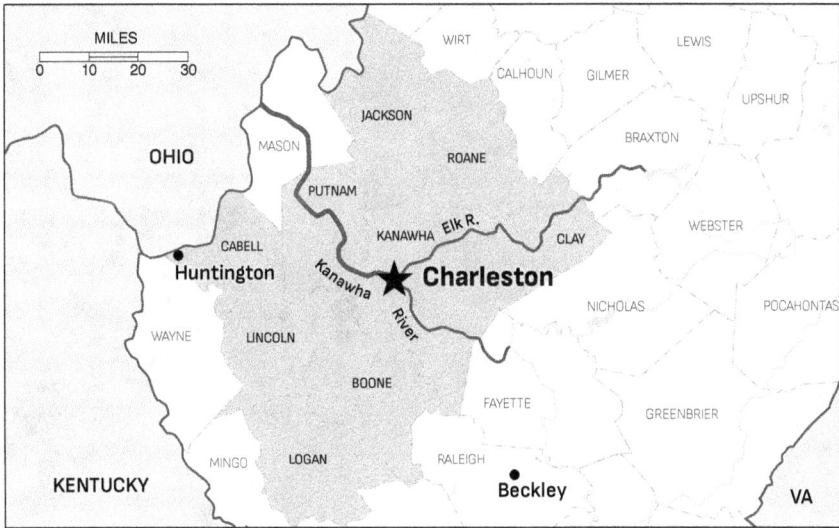

Overview of the area served by the West Virginia American Water treatment plant on the Elk River. (Map by Than Saffel, based on data from West Virginia Water Claims.)

Map depicting the Do Not Use zone. (Map by Than Saffel, based on data from West Virginia American Water.)

Freedom Industries tank farm on the Elk River, Charleston, West Virginia, shortly after the spill. (Courtesy of Wesley A. Kuemmel.)

Empty bottled water shelves in the Charleston area during the early days of the water crisis. (Photo by Elizabeth Campbell.)

For those businesses that remained open, bottled water was offered to customers and used in food preparation. (Photo by Elizabeth Campbell.)

Members of the West Virginia water crisis project seminar meeting at the South Charleston graduate campus of Marshall University. (Courtesy of Wesley A. Kuemmel.)

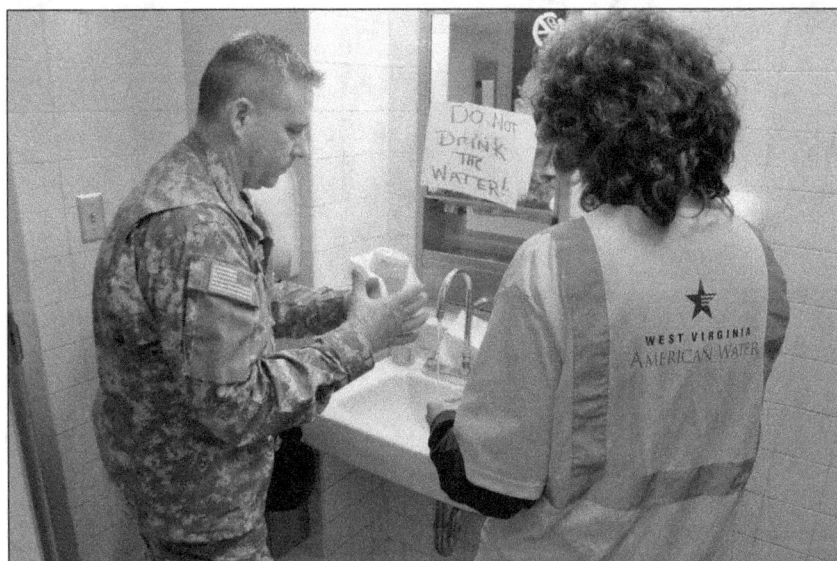

A West Virginia National Guard member draws a water sample in the Kanawha Valley to determine levels of contamination during Operation Elk River Spill. (West Virginia National Guard.)

Effluent from the Union Carbide chemical plant flows into the Kanawha River in 1973. (Photo by Harry Schaefer for the U.S. Environmental Protection Agency.)

Following are step-by-step procedures customers can use to flush their plumbing system. To protect the health and safety of our communities, we recommend that you read carefully and follow the steps for flushing. Thank you for your cooperation. **NOTE: After flushing, your water filters need to be replaced. If you have any point of entry water treatment system such as a water softener or filter, please refer to "How to Flush Plumbing Appliances and Faucets."**

West Virginia American Water will be offering residential customers a credit of 1000 gallons, which is more than what will likely be required to flush the average residential home. The average residential customer uses approximately 3,300 gallons per month.

## How to flush your plumbing system

**Please complete these steps in the order set out below. Finish each step completely before moving on to the next step.**

**1. Flush ALL hot water taps for 15 minutes**

Begin the flushing procedure by opening the hot water taps in your bathroom(s). Open ALL hot water lavatory (sink) fixtures, hot water bath fixtures, and any other hot water fixtures, such as kitchens, wet bars, etc. **Run these hot water fixtures for at least 15 minutes. Shut water off after 15 minutes.** After you have flushed each hot water faucet for 15 minutes, your hot water heater will be safe for use.

**2. Flush ALL cold water taps for five minutes**

Once the hot water tank and hot water piping have been flushed, open ALL of the cold water fixtures, flush each toilet at least one time. **Run these cold water fixtures for at least five minutes. Shut water off after five minutes.** This does include the water in your refrigerator water dispenser.

**3. Flush ALL remaining faucets and appliances**

(Before starting step 3, please see **How to Flush Plumbing Appliances and Faucets** for more information.) Open any remaining fixtures such as hose bibs, external faucets or fixtures not used for drinking for at least five minutes to finish the plumbing system flushing. Take additional steps to remove water from other appliances. See **How to Flush Plumbing Appliances and Faucets** for more information. This includes:

- Ice makers
- Dishwashers
- Washing machine
- Humidifiers
- Continuous Positive Airway Pressure (CPAP)
- Oral, medical or health care devices
- Baby formula, food and drinks made with water during DO NOT USE
- Water filters
- Water softeners
- Reverse osmosis units

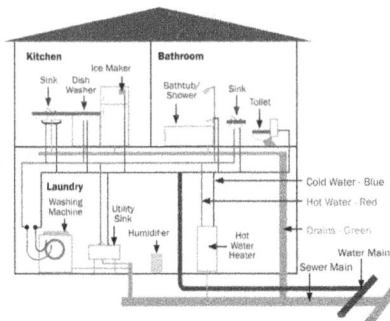

**Any lingering smell, which is expected, is not a health issue.**

For more information: Please contact our 24-Hour Customer Service Center at 1-800-685-8660 or visit our website at **www.westvirginiaamwater.com**.

Page 1 of 3

West Virginia American Water website notice to customers for "flushing" their plumbing. (West Virginia American Water.)

Cleanup begins at the Freedom Industries tank farm on the Elk River, Charleston, West Virginia. (Courtesy of Wesley A. Kuemmel.)

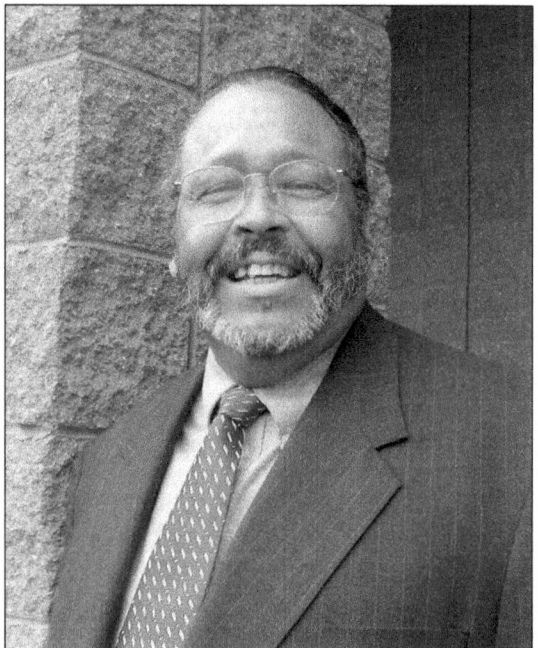

Paul Gilmer Jr., local business owner, reflected on how the crisis focused his commitment to place: "There have always been many different things that happen [here] that make you think about, you know, 'Would I be better off in North Carolina? Or would I be better off in Ohio?' I always think about that. But there's nothing that's made me pack a suitcase. I'm committed here, my family's roots are down here. I've got my business roots down here." (Courtesy of Schoenbaum Family Center.)

Carla McClure didn't think of herself as an activist before the spill. But after the spill, she said, "I started hearing about people who wanted to get together and do something—though at that point nobody was sure yet exactly what, other than let's see what's happening." (Courtesy of Carla McClure.)

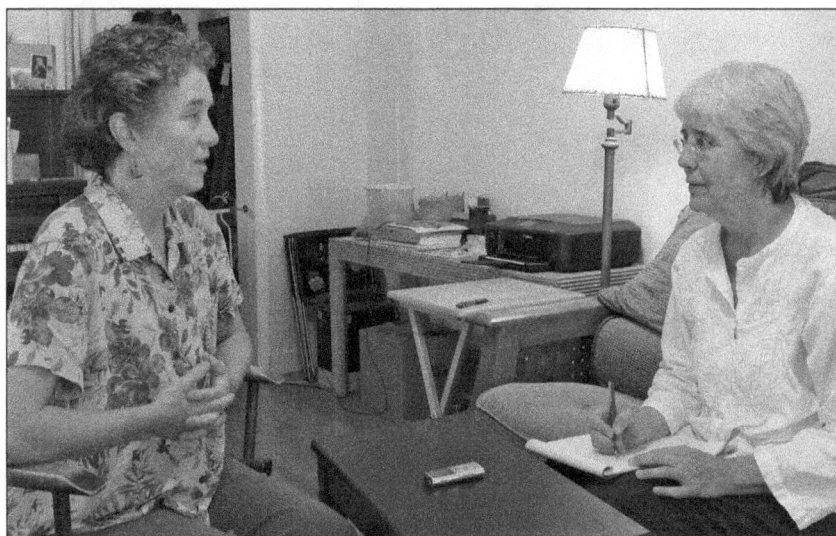

Becky Park, climate lobbyist, speaking to Trish Hatfield (*right*), called the crisis a "positive experience" because "finally the public was thinking about what was coming out of their taps." (Photo by Jim Hatfield.)

Keith Redmond, shuttle driver and bellman for the Marriott Corporation, thought the crisis was devastating. The hotel was closed for five days. "Our waitresses live from payday to payday, or day to day on tips. It was tough." (Photo by Trish Hatfield.)

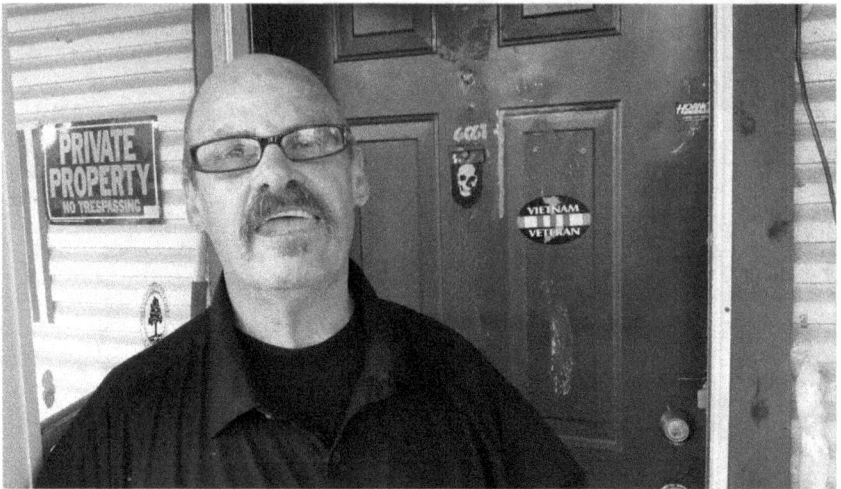

Kenneth Mize, veteran of the Vietnam War, was nervous about being interviewed but now is pleased to be part of this project. "There's no reason we should have this experience. Not in America. Not in this day and age." (Photo by Trish Hatfield.)

Paul Epstein, musician and teacher, turned to his musical roots to address the crisis. Paul said, "I realized that, as a musician, with a lot of ties to musicians and artists who are often being asked to provide entertainment or art to support environmental issues to help raise money . . . I thought, 'Well maybe there's a way I could use my expertise to help.'" (Photo by Al Peery.)

Marlene Price, who works for the West Virginia Department of Education, turned her attention to those who might not be in a position to find water right away. "I had bottled water at home because I keep it for travel," she said. "I had a couple of cases, but what about the people who didn't? And I became concerned about the elderly, the people who couldn't get out to get some water. And poor people who can't afford it. Immediately my family was calling and saying, 'Who's checking on so-and-so? Do we need to go by there? Let's run up and see if we can get water from Walmart or Sam's or something, [so] that we can buy several cases and just give it to people so they'll have it.'" (Photo by Trish Hatfield.)

Part II    **On
Place:
To Stay
or Not
to
Stay**

# CHAPTER 5

# Blues BBQ

Jay Thomas

In January 2014 I owned Blues BBQ in Charleston. On January 9, the spill's "official" beginning, my wife, Honor, and I were driving our daughter, Daisy, to Shepherd University in Shepherdstown, West Virginia. About 7:30 p.m. or so, Vincent, a Blues BBQ employee, called Honor's cell to tell us that there was some sort of water problem and that we would have to shut down immediately. I told Vince to wait for something from the Health Department before doing anything. He called back two minutes later and informed us that the Health Department had just called and said that a chemical had leaked into the water supply and that we must close until further notice. My driving became erratic as I wondered what the hell was going on. I asked Honor to call Julie, a bartender at another restaurant, Bruno's, which I co-owned at the time. Julie repeated what Vince had said. Honor and I, hardened toxic warriors that we are (we've lived in Charleston for many years), figured it was like a power outage or gas leak and that it would be over by morning.

We arrived in Shepherdstown, had some dinner, and talked about what we didn't know. I felt no heightened sense of alarm at the time, though, just irritation from the lack of concrete information and an evening of no restaurant or bar revenue. Be that as it may, it was getting late, and we were exhausted from the long, post-workday drive. We checked into the hotel, slept soundly, and prepared for Daisy's college initiation the next day.

The next day, as we navigated orientation, class registration, dorm assignments, and all other things college, we talked briefly with some other parents, who shared some information seeping out from Charleston, but no one was clear about anything. Several times I broke away from the day's busy activities to talk to Julie, who reported the same from the powers that be: stay closed until further notice.

We were in Shepherdstown for another day, and as Honor and Daisy were off shopping for school supplies, I took the opportunity to get a haircut (we had a dinner engagement later with my in-laws). Strolling around downtown I spied a somewhat rough-looking hair salon that looked like me. Sure enough, as I walked in, I immediately noticed a guy considerably younger than me, but with the same type of haircut, who was sweeping around one of the two barber chairs. He brushed off the seat, I sat down, we conferred on specifics, he started my haircut—and I finally learned more about what was going on in Charleston.

Ben the barber, it turned out, knew more about what was happening in Charleston than I did. He had a television in front of the chair, with news reporting information as it came in. Authorities had not publicly identified the chemical as MCHM yet, but the reports were that residents in nine counties were told not to use the water except for flushing toilets. I realized then and there that this was a much bigger problem than I had imagined. For the first time, I experienced a sense of fear.

Ben was originally from western Pennsylvania, not far from Morgantown. "Yeah, to be honest," said Ben, "I'm not surprised. We've had water problems in Pennsylvania forever, it seems like. Rivers, mostly. Dead fish by the hundreds floating downstream from the coal plants. People getting sick before anyone realized there was a problem with the water. You guys are what, in the Chemical Valley? I'm surprised something like this hasn't happened before."

Ben also knew more about the crisis than what those on the television seemed to know. In addition to being a barber, Ben was also an environmental science graduate. As he cut my hair, he went on to talk about how at least half of environmental students change majors after the first year, especially once they realized the futility in trying to change the way most industry does business. Ben frequently paused his work to elaborate and answer my questions. More than a few times, he used a nearby dry-eraser board to draw chemical compounds for me, talking about isotopes and stuff I hadn't studied in thirty-some years. I learned about "dilution to treat pollution" and other industry approaches to chemical contamination as he cut my hair during what were now breaks in our conversation. He predicted (correctly) that the company responsible would file for bankruptcy almost immediately. And lots of other stuff. To say the least, that was the longest haircut I've ever had. I gave him a generous tip, and I left with my fear starting to turn to anger.

A favorite movie of mine comes to mind here: *Coming Home*. Released in the late 1970s and starring Jon Voight, Jane Fonda, and Bruce Dern, the film revolves around two men who go off to fight in Vietnam. They come home changed forever, physically, emotionally, and mentally. I've watched it several times over the years, and it never ceases to affect me. I would never presume to know or understand what Vietnam veterans went through; that's not my point here. But the next day when we prepared to head back home to Charleston, I felt, as portrayed in that movie, like I was a changed man. I could never look at the Kanawha Valley and Charleston the same way again. Talking to folks in and around Shepherdstown, I was embarrassed. People were polite, but I felt disdain for southern West Virginia. More than a few folks said they weren't surprised, what with our living in Chemical Valley. Several asked if we might move. It was the first time in my life I hesitated to say I was from Charleston.

As we drove home, down Interstate 79—and as we passed industrial sites along the river—we listened on West Virginia Public Radio to a former Charleston resident, Eric Waggoner, read his essay "Elemental."[1] Waggoner's piece, which went viral soon after the spill, expresses a complex and multidimensional anger with the whole situation. But, essentially, he said everyone involved could go to hell, from the coal industry, chemical factories, local and state governments, and even good ol' West Virginians themselves. "Since my own relatives worked in the coal mines," he said, "and I can therefore play the Family Card, the one that trumps everything around here: To hell with all of my fellow West Virginians who bought so deeply into the idea of avoidable personal risk and constant sacrifice as an honorable condition under which to live, that they turned that condition into a culture of perverted, twisted pride and self-righteousness, to be celebrated and defended against outsiders."

No one was blameless. Them, us, everyone deserved blame, whether they were intentionally ruining the state's environment or too apathetic to do anything about it. "I'm not an eco-warrior or a Luddite," Waggoner continued, "and I'm not anti-business or even anti-industry. But for years, I've watched from inside and out while the place I grew up in, the place where many people I love still live, got sold out and scorched and plowed under and poisoned and filled with smoke."

Listening, and while looking at the industrial tanks along the river as we passed by, I thought I should feel more of Waggoner's anger myself. But I found myself not feeling so much angry as feeling so extremely sad. I didn't want to live here anymore. I'm originally from here, grew up in South Charleston but left and then moved back thirteen years ago. Thinking this was a great place to raise a family, I brought my family here. And at that moment, listening to Waggoner, I felt that I'd betrayed them. If the government won't protect its own people, I thought, how could we continue to live and work here?

# Keeping My Restaurants Open

I came to this project as a student of the graduate humanities program. But it was in a somewhat unusual way. I was first an interviewee, interviewed for the project by Cat Pleska, who wanted to know about how my restaurants weathered the spill.

In any case, I dealt with the crisis as best I could, bringing back home from Shepherdstown a few hundred gallons of water to support our home, businesses, friends, and family. After I got home, for the first day or so, I was bewildered about what happened and, to be honest, wasn't sure exactly what we'd do. At first, we weren't getting any solid information; we just knew that something had leaked into the water system. But once I started finding out just what had happened, I got more and more angry. You can't run a business, a restaurant (or two, in my case) without depending on clean water in your taps. I was angry about the lack of information but even angrier about the lack of regulation, the almost nonexistence of facility inspection, that MCHM had leaked into the system in the first place.

Getting our restaurants back open was a real challenge. We couldn't use the tap water, of course, so the Health Department called and told us that to reopen we needed to come up with a plan to find and use water that we could cook and clean with. Then they would come and inspect us based on that plan. It took some time for the Health Department to make it around to all the restaurants in the Charleston area. While we waited for the inspection, among other preparations, we brought in bottled water to cook and clean with, set up handwashing sinks with bottled water, and added filters to our ice machine (to be used once the tap water was potable again). We passed the inspection and reopened after being closed for eight days. We lost a lot of revenue, of course, and it was a difficult time for my family and me, but it was especially hard on our employees.

Running a restaurant is tough business. Something like this makes it even tougher, especially when employees lose wages (which include tips) they depend on. That's their income, and they were out of work for more than a week. Many of us are close friends, and we've worked together for a long time. So I did my best to help support them during that time. Several applied for the help offered by the state government to recoup lost wages, but the state was pretty slow to respond. Though some did get help, they never made up those wages. Even so, our employees were great during this time. They were dealing with this in their own homes, but they did everything they could to ensure we reopened.

Eventually the state deemed the tap water potable again. But I was still very skeptical that the water coming out of our taps was actually safe to drink. We didn't use water from the tap at home or at the restaurants for a very long time. I just didn't trust what we were being told. And I didn't trust our government to tell us the truth about the water we depended on for survival. To be honest, my trust will never be regained completely.

## Time Has Passed

After the crisis, many of the people I knew were upset. They said that something needed to be done and made a movement toward getting people together. And I thought it was great. We went down to the Unitarian church for a meeting of concerned citizens, and there were several people from the legislature there, several people from environmental advocacy groups. You signed on, they got your email, and we communicated, and then they had another meeting. They expected about a hundred people, and about 250 people showed up. So you could tell that this might be a blessing in disguise is the way I'm looking at it now, to get people activated, to get people away from the idea, "Well, this is just the way it is." But I started going to local meetings with environmental activists and local business people, folks trying to change the way we live here.

Now, I think, people know what's going on out there, and they want answers. They're not going to take things for granted anymore. They want clean water and clean air. And I'm still involved, more than ever. Like this oral history project: I got involved as a student of the graduate humanities program as the project developed around two graduate seminars (described in the introduction). But to me my involvement is more than just a graduate seminar. It's about life *here*, in Charleston. Regardless of what the outside world or greedy industrialists or apathetic West Virginians might think, we matter. And thank God not all environmental science majors drop out. We need them.

## Note

1. For a recording of Eric Waggoner reading his entire essay, see Scott Finn, "'To Hell with You'—A West Virginian's Raw Response to Water Crisis Goes Viral," West Virginia Public Broadcasting, January 18, 2014, accessed September 24, 2017, http://wvpublic.org/post/hell-you-west-virginians-raw-response-water-crisis-goes -viral.

# CHAPTER 6

# Citizen Response: On Leaving and Staying

Cat Pleska and Joshua Mills

"Probably this could happen about anywhere," said Anita Edmonds.[1] "I don't feel like it's any more dangerous to live here than it would [be] anywhere else. I kinda feel like it would be safer to live here." Edmonds lives in the small town of Hamlin, West Virginia, where, for many decades, everyone has known everyone else and family names have repeated down through the last 150 years. Names like Stowers, Adkins—and Atkins—Jenkins, and a handful of others. And no doubt *Edmonds* is familiar, too. Anita's father is from Lincoln County, and her mother has lived there all her life. So has Anita. To her it's a given that after the chemical spill, first responders went from door to door in this small village, checking on the elderly to see if they needed water or anything else.

When asked if the water spill and crisis prompted her to change her attitude about her state and her town, she said it did not. It would take a major disaster for Anita and her family to leave their beloved area, she said. It's where home, friends, and family, people she has always trusted, have always been.

## Native West Virginians: On Us

As it happens, we are also native West Virginians, and we're both from small, rural towns.

I (Cat) am a seventh-generation West Virginian. My people were mostly agrarian until mid-twentieth century, when the men in the family worked in industry, such as the gas company and aluminum manufacturing. I grew up in Hurricane, in Putnam County, with a population around three thousand. Living in a small town also afforded me a feeling of safety, where I knew most everyone and could feel a connection, that I was never truly alone or without support.

Still, Putnam County, where I still live, was not among those struggling with the aftermath of the chemical spill and resulting water crisis. This escape from disaster didn't seem unusual to me on one level. Much of the state's history has experienced disaster after disaster because of extractive industrial accidents (mining and timbering, mostly), and Putnam County has emerged largely unscathed as it has remained mostly agrarian and then residential. It turns out, however, disasters were close by—either covered up or not discovered for decades—that did affect us in this quiet county. Kanawha Valley, near where I live, is known as Chemical Valley. The chemical companies polluted our rivers and air. I once received a class action lawsuit letter because I had lived most of my life in the airflow path of a dioxin contamination.

Still, though, a chemical spill contaminating the water supply so close to me and that affected nearly three hundred thousand people was a shock. Knowing that long-standing pollution and disasters were not likely to be addressed or changed by our state political leaders, I wondered, unlike Anita, when policies would change and we could all truly feel safe. I said to anyone who'd listen, not for the first time: How many sick or injured or dead people does it take until our leaders make changes and say "enough"?

As the hours passed after the first announcement about the spill, information was sketchy at best and continued to be so for days (a not-uncommon reaction from our political leaders). Eventually, friends who had no access to water for bathing accepted my invitation to shower and wash clothes at my house. When they showed up, disheveled, with towel and soap in hand, it felt surreal for them and for me. While their bodies were cleansed, the hearts and minds were not, as everyone wondered when the ordeal would end.

Later in the year of the spill, at the invitation of West Virginia Center on Budget and Policy via Eric, I signed on to conduct oral history interviews concerning the water crisis. Once our group of interviewers decided on the demographics of our interviewees, we fanned out to question those who had directly experienced the chemical spill. By late August, I had completed my round of interviews.

---

I (Joshua) grew up in West Virginia, too. I've spent much time around and in water, and I am certain I will always have some amount of West Virginia's water soaked into my bones. I have spent countless hours catching crawdads in Mountain Mama's creeks, fishing in her rivers, swimming in her lakes, and enjoying the coolness brought on by a summertime rain.

I grew up in Wayne, a small rural town, and though it is not the place it once was to me, I still hold the fondest of memories of my childhood. A young boy could not ask for a better playground. My childhood home was surrounded by several hundred acres of woodlands my grandparents owned. When my brother and I were not hunting or playing in the woods, we'd find our way to the baseball or football field. We seemed to be always playing, even in the pew at our small church. I had everything I needed and wanted as a child.

Along with childhood play, another common thread running through my childhood memories was coal. I loved trains as a boy, and lucky for me, a coal train ran twice a day right through the heart of Wayne. And it also ran just a few feet from the dugout at our ball field. I remember many times during our Little League games, the train would pass and drown out our cheers as we played. Many hours of my life I spent watching coal trains go by, stopped at crossings, listening to the screech of the wheels and reading the graffiti on the side of the cars. Even at my high school, some of the rooms were so close to the tracks we could feel the room shake as the train passed, whistle blowing and drowning out our teachers' instruction.

I never really considered how complex, and at times difficult, the relationship between my hometown and coal could be. The coal train was like the water: it was just something that was there, always had been, always would be. And like water, I took it for granted. That is, until the last train ran through our town in 2015 after the mine shut down.

When I initially heard of the spill, to be honest, I was not very concerned (I lived in a region unaffected by the spill). As the crisis continued, though, I began to sympathize more and more with those who were still without water. Still, though, because I was not directly affected, I gave the crisis little serious thought—that is, until Eric asked me to be part of a graduate seminar that would make a radio documentary about the crisis. Through that seminar, I acquired a new appreciation for the water I had played in and drank all my life and realized the precarious state of our water infrastructure, not just in West Virginia but everywhere. As a result of that seminar, I came to work with Cat on our part of this book project.

## On Rural Experience

The chemical spill affected individuals living in both urban and rural settings, and individuals within these respective settings faced many of the same challenges in their search for clean water. Individuals affected by the spill outside city limits, however, experienced slightly different challenges, and their attitudes reflected their different experience.

In a rural area, key differences from urban areas include distance from infrastructure such as government centers, health facilities, (at times) close neighbors, work sites, and sometimes safe roads. Particularly in West Virginia, with its mountainous terrain, simply traveling to larger city centers can mean very long drives.

The toxic chemical that snaked its way down the Elk River and eventually into the waterlines of hundreds of thousands of West Virginians was unbiased in its attack. It intruded into homes, government buildings, restaurants, and schools. It flowed from the marble sinks of the wealthy just the same as it did from a rusty faucet used for watering animals. No living thing in the affected region was exempt from the tainted water, and everyone faced the same dilemma: "Where can I get clean water?"

In our oral history project, many people living in rural areas, in the periphery of the affected zone, actually found gaining access to clean water a bit easier than their urban counterparts. Belinda Vance, who Cat interviewed, is an example. She lives in a rural area of Cabell County. She's a retired schoolteacher who taught band at Hurricane Middle School in nearby Hurricane for several years. She lives on fifty acres in a home she, her husband, and two sons built with a number of barns and sheds to house horses, goats, chickens, cats, ducks, dogs, and, at one point, a mule named Sister Sarah.

Because of the spill, Belinda and her family had to also worry about farm animals and pets. She reported that many friends, neighbors, and family members reached out to her and her family, offering them use of their water. One of the local churches also opened their facilities for those affected by the spill. As for her animals, she said her son was able to contact a friend in the nearby town of Milton, an area unaffected by the spill, who had a large mobile water tank they used to bring in water for the livestock.

Unlike those living in urban areas, many rural residents had to struggle with uncertainty caused by the delay of accurate information coming out of Charleston. "We had been watching the TV, and the alerts came on that there had been a spill, and it was affecting Kanawha County," Belinda remembered.[2] "Later it came on that it was affecting Putnam County." So this left Belinda and her family initially uncertain about whether their water had been affected. "We thought, 'Okay, that's us,' but they hadn't said Culloden [an area in both Cabell and Putnam Counties]. They said Putnam County, not Cabell and not Culloden. But we thought, 'No we're in that.'" So to clear up the confusion, Belinda's husband "went out to the [Culloden] fire department and talked to the fire chief, and said, 'This also includes us, doesn't it?' and he [the chief] said, 'No, our water comes from Huntington.' . . . That was about 5:00 p.m., and at

8:00 p.m. it came on the news [that] Culloden is in it as well." Though solid information about whether they could use their water traveled slowly to their home on the periphery of the affected area, Belinda and her family did not use the water in any case.

Similar to Belinda's experience, Anita Edmonds also reported that she was never that worried about having access to clean water because of nearby friends and family members who had well water not contaminated by the spill. The local fire department and Red Cross also provided clean water. But this didn't mean that the crisis caused less anxiety or stress in the context of rural networks of kin and neighbors. As Belinda had, Anita and her family faced uncertainty from delayed and relevant information. "It really worried me," Anita said, "because if it happened that easily, how often has it happened and we didn't know about it? We never smelled it. [Not having clean water] didn't worry me as much as the *not knowing*, and then finding out that maybe it had happened sooner than what they had said, and maybe we had already been drinking it. I was concerned because you don't know what the chemical could do if you had been drinking it."

Accurate information about the spill was not the only thing that slowly crept out of Charleston. Because of the distance of rural areas from the water company, the chemical itself reached many rural residents much later. While many residents in Charleston may have sensed something was wrong by the odor in the air and the taste of the water before actual warnings were given, those living in rural areas "didn't smell it for weeks," said Belinda Vance. "The whole deal was almost over with before we ever smelled it. Actually, they had even given the okay to use the water again before we ever smelled it."

Anita and her family, who live around thirty-five miles from Charleston, had a similar experience. "Our water did not smell. It did not even smell until we started flushing our pipes. We would even turn it on to smell it because everyone said it smelled, but we never smelled it." Other residents reported similar experiences, that it wasn't until they began flushing their pipes, as instructed by state officials, that the smell of the chemical became more prominent. "When they said flush your lines," Anita recalled, "we did. But then they said you have to do it again. It became really aggravating because you really didn't know. And that's when it really started smelling. Where they opened the water line on our street it ran into a creek . . . and you could smell it when you walked out our door." As the flushing process became more involved, "that's when you could really smell it," Anita continued. "The odor lasted for days outside. Then you could go a week where you didn't smell it. Then all of a sudden you would smell it again. It was really kind of strange. It was really another reason we questioned if we could really use the water yet."

## On Urban Experience

Paula Clendenin, the Charleston artist, has an apartment a block away from the state capitol. The day after the crisis officially began, and when it was clear that she and her cats would need water for some time, Paula set out to find water. She had heard most of the local stores were sold out of bottled water, so she left the city in her search. "Everybody was running for water," she remembered.[3] "And I thought, 'I'll go up to Walmart in Rand'"—a small community about ten miles up the Kanawha River from Charleston. But when she arrived there, she found it barren of water. So, like many others here, she adapted. "I got stuff I usually don't drink," she said, "like bottled pop . . . just to have fluids." But, of course, bottled soda was useless for things like washing dishes or bathing or hydrating cats. So her search for water continued.

Eventually, that evening, Paula got water from the National Guard, "on the boulevard [near the state capitol], with these big searchlights" illuminating the water trucks. It was a surreal scene, she reported. "Hundreds and hundreds of people," she remembered, "just walking towards the light, across the lawn [of the capitol] and down the street to get in line and get water. But no one was talking. It was so quiet and eerie." And it was at that point that Paula began to realize that this was serious. "You start seeing the National Guard," she said, "and it's like, 'God, this really is a disaster.'"

Many fellow Charlestonians faced similar struggles. Kenneth Mize, for instance (introduced in Trish's chapter 4), had no car, and thus had no way to drive to the superstores, or even to many of the water trucks that were placed throughout the area. Sue Brookshire, a sixty-seven-year-old retired elementary school teacher who lives in Charleston, also had difficulty finding water. "Any time they tell you not to drink the water, including the entire three hundred thousand people, you know that's a serious sort of thing," she said.[4] "I normally am not a bottled water drinker. But then I started watching the news, and there was a rush on the stores to get all the water. . . . People had rolled out grocery carts filled with bottled water." It didn't take long for Sue to also realize that this "was a serious problem." So, she said, "rather than running all over Charleston, we decided to go out of town, which turned into quite an experience," much more difficult than she had imagined. "The next day my good friend . . . called a friend in Lexington [Kentucky] and we drove halfway to Lexington. She and her son met us with a carload of water."

Despite having a carload of bottled water, out of caution Sue and her friend stopped in Ashland, Kentucky, nearly seventy miles from Charleston, to buy what more they could. They went to a Walmart there, only to find barren shelves where the bottled water should have been. She was told that the Charleston Fire Department had been there earlier and taken all the water. Fortunately, another Walmart in the area still had water in stock. So she and her friend were able to purchase more bottled water "to make sure we had it." Sue also said they bought baby wipes, which they used to bathe themselves over the next three and a half weeks.

## To Leave or Not to Leave

For many, the water crisis was the final straw in what they saw as a long history of problems in our state. In fact, many natives have left the state in an exodus that has occurred, actually, over many decades. They've left for a variety of reasons, not least of which is the dearth of employment. Add to this how large industrial corporations have driven many rural and urban residents alike away from their homes with similar environmental contaminations. Indeed, many connected the spill to other industry-related disasters that have occurred throughout West Virginia's history and blamed the political leadership for not protecting residents from negligent corporate industries. Many are all too aware, sadly, that our politicians here have a tradition of favoring big industry—which operate in many counties as absentee corporations—to the detriment of our health and well-being. Political leaders cite employment and tax revenue as the basis for their decisions. We need both here, to be sure. But, as is also well known, a more equitable balance between outside and inside interests is still a distant dream for most West Virginians.

Several of our interviewees had lots to say about these power imbalances. Matthew Chesebrough, for example, was a Charleston resident and student at West Virginia State University at the time of the spill. He quickly identified himself as being among those who blame the state's leadership for not providing information and speed in protecting the state's citizens. Matthew is a young professional, smart, quick witted, and driven. He is a millennial who is already tired of politicians' and community leaders' excuses. When the spill happened, he felt disgusted and frustrated with the fact that it happened in the first place; the ongoing lack of response and information made him declare that he might leave the state before he had planned.

"It's not the first time this has happened," Matthew pointed out, "and it's not going to be the last time."[5] With frustration in his voice, he was pretty emphatic about this. "Nothing has changed. If you look at West Virginia's history, just over the last five or six years there's been numerous spills of chemicals and coal slurries. This state is not well known for its environmental concern. It was just like 'okay, it's another one.'"

Similar to others, Matthew reported that at the beginning of crisis, his main source of frustration was the lack of clear information about what exactly was happening and how to cope. "We just kind of dealt with it in the beginning," he recalled. "There wasn't any information. That was one of the most difficult aspects of the whole event, that there wasn't any information as to what was going on. . . . The left hand doesn't know what the right hand is doing. Every news conference or announcement was just the same generic information, rehashed or reworded."

As the crisis wore on, Matthew's frustration turned to anger. He felt very strongly that corporate industry and corrupt politicians were to blame. "Once again, it's same ole, same ole for this state. . . . Just another reminder of how much business is put in front of people in this state." When I (Cat) asked him about it, he seemed to have little hope that anything would actually change as a result of the spill. "At the end of the day," he said, "I'm all for this hippie mentality of 'let's protest,' 'let's sign little pieces of paper saying how we're not happy,' and candlelight vigils. . . . [But] it didn't change a damn thing. And if you believe it changed anything, you're mistaken. It's business as usual."

Though business as usual, the spill was a breaking point for many living here. And Matthew was no exception. "The number one thing that affects us is that we're leaving," he said, reflecting on the spill's effect on Charleston and the surrounding areas. "We had planned to stay in the city and buy a house and build a family here, but we've decided that we're leaving. . . . We want out. There is nothing that this state can do to persuade us not to leave." Though many others felt this way, he argued, this was just one of many ongoing problems keeping people from living in West Virginia.

Matthew and his girlfriend are "tied in to the younger professional community [in] the city," a population that Matthew argues will leave and choose elsewhere as long as the options for "modern living" in West Virginia continue to decline. We have a "brain drain of your professional upper-middle-class, lower-upper-class individuals," he continued. "It's more efficient to just get the hell out of the city, get the hell out of the state, than to stay and try to build a life here, to build businesses, to build a family here. . . . Nothing will change."

Others we interviewed had much to say about this issue. Paula Clendenin was also frustrated with the lack of official action. She was then and remains now, like many natives, angry that the atrocities to the land and the people have continued apace for the last 150 years. Paula echoed many of Matthew's sentiments regarding the crisis, particularly corrupt industry and politicians. "Right now," she said, "I'm into the stage of fight or flight. . . . Our government and our representatives and our legislature are not for the people. It's like we're a colony. . . . It's 'for the people, by the people,' but it's not [that way] in West Virginia. It's *for* the corporations. It's *for* the coal industry, *for* the fracking. . . . It's King Coal that's held this state for so long. And it's killing us. And the people who we are paying to protect us don't care."

Paula's sense of hopelessness led her to the difficult decision of finally deciding whether or not she should leave the state. "I feel like I'm in an abusive relationship with West Virginia," she said. "I love it but it keeps abusing me. . . . I keep thinking that if I love it better it will get better." After some reflection, she seemed resolute about whether it will change for the better: "I don't think it can. . . . There's been decades of being in a coal colony, and I don't think you can change it." With the issue of staying or leaving the state in mind, the question for Paula, then, became: "Do I want to be a secondary citizen in this corporate world?"

---

Such powerful sentiments, we should note, were expressed alongside another set of powerful sentiments about *not* leaving. Though most others affected by the spill shared similar frustrations with industry and political leadership, they expressed these frustrations in ways that reinforced their connection to place and family. This was particularly true for those who lived in rural areas, like Anita and Belinda. Anita, for instance, did not hesitate to express her frustration with the water company and government officials. "They didn't know how to handle it," she said. "I don't understand why you would not have something in place to keep it from getting in the water." But, at the same time, she felt that "you gotta do what you gotta do. That's just the way I looked at it." All told, she said, she felt safer where she lives—in a rural area—than she would elsewhere and even noted that a friend of hers moved from Charleston to the more rural Lincoln County as a result of the spill.

Belinda—whose property is at the end of the American Water Company's line in Culloden, the only portion of a line the company has in Cabell County— echoed Anita's sentiments. They did "pretty much what I expected them to do—pass the buck, fabricate things," she pointed out. But she also noted that she "wasn't all torn up about it. . . . You just have to accept it for what it was and go on and just deal with it." Being reared in a family that raised cattle, she's certain that a rural setting suits her best. Her husband worked in the chemical industry for thirty-three years, and she said the spill was "not a surprise" given the numerous chemicals used in the area. Current regulations on the chemical industry, she said, are enough to prevent disasters like the water crisis, but such regulations are poorly enforced. So, like Anita, her sentiments about living in West Virginia were unchanged as a result of the spill. "Things happen everywhere," she noted, and "we came out unscathed as far as we know. I just don't get all torn up about stuff. . . . Not all chemicals are bad."

---

This issue—to stay in or to leave West Virginia—is, we want to note, complicated and multidimensional. On the one hand, people here often talk about their love of place and family, their connection to the land, to community, and to each other. But on the other hand, they also talk about their struggle or inability to stay here because of, for example, the lack of economic opportunities, low education levels, or corporate and government corruption. Events like this chemical spill amplify the intensity of this ongoing conversation. And redouble its complexities.

"I'm a native," Paula reminded us. "I did the native thing that I had to get the hell out of here as soon as I could [when I was young]. But I came home for family reasons. . . . I like the people." And, in that same breath she was quick to add, "I love West Virginia," pointing out that when she moved back, "things opened up for me. . . . It's affordable for the lifestyle I want. I have a big studio I couldn't afford anywhere else." To consider leaving, then, is actually a complicated prospect for Paula.

These multidimensional conflicts, of course, are rooted in a deep historical and cultural bond with West Virginia. Many, many native West Virginians can trace ancestors back well before West Virginia was even a state. Family ties thus run deep. We have forefathers buried on rural ridgetops and their descendants living in the same hollows and valleys that they first settled. I (Joshua) personally have family members who live in the original cabin built by my third great-grandfather. But as Paula once did and many others still do, I must leave West Virginia to find real professional employment opportunities. I do hope I can return one day.

I (Cat) sympathize with Joshua's sentiments. My roots run deep, too. I am the great-great-great-great-granddaughter of a veteran of the War of 1812, a man who had, at that time, already settled in what would become West Virginia many years later. I have the original land grant document granting land to my great-great-great-grandfather in Putnam County two months before West Virginia became its own state. It seems more than ancestry to me; it almost seems as if I have genetic memory of all who came before me and loved this state as I do.

Many of those we interviewed noted family ties like these. Take Carla McClure, who at the time of her interview lived in a section of Saint Albans affected by the chemical spill. "This is my hometown," she said.[6] "I never want to leave it, and this is my home state. I've left before and wanted to come back. I had lived in Utah and North Carolina, and both times was so homesick I couldn't wait to get back." Or take Charleston resident, Julian Martin, a retired high school chemistry teacher and active environmentalist, who reminded us that "I'm eighth-generation West Virginian. My people came over around 1800 from Virginia. I was born down the river from where they originally settled on the Coal River. . . . My dad was a coal miner. My grandpa was a coal miner. My grandfather was in the battle of Blair Mountain, . . . he and his brother, on the side of the United Mine Workers. . . . My son has two great-grandfathers who were at the Battle of Blair Mountain."[7]

Again and again, people were quick to remind us of their strong ties to the state, how it was a beautiful place to live or how they were connected to the culture of the region. For those like Paula, though, talking about their love of the state often came with a "but." To be sure, many West Virginians have a complicated relationship with their home state, an enigma perhaps captured best by Matthew Chesebrough: "I love my state, but my state is very sick right now."

## On Staying

West Virginia, of course, is part of the much larger region of Appalachia, which covers parts of thirteen states. West Virginia is the only state wholly within Appalachia. Much of this region is often characterized by poverty (particularly in the more mountainous areas, which flow up the region's middle from Georgia to Maine); is mostly working class; and in areas somewhat underdeveloped, is often dominated by large industry (most of these companies absentee). The people of Appalachia (around twenty-nine million) are often imagined as stoic and resilient and inordinately dedicated lovers of its beautiful, challenging landscape.

To a certain extent, these characterizations—stereotypes to many—are true: many insiders, for example, also imagine themselves as stoic, resilient, and inordinately dedicated to place. It's often what keeps us here in times like this. But that resilience can also be a double-edged sword, especially when it is embraced uncritically, wrapped in a cloak of stoicism that at times enthusiastically accepts the difficulties of living here as part of the price we pay to *be* here. It's not hard to understand, then, why many end up buying into, as Jay Thomas reminds us (in chapter 5), and as Eric Waggoner eloquently put it: "the idea of avoidable personal risk and constant sacrifice as an honorable condition under which to live . . . a culture of perverted, twisted pride and self-righteousness, to be celebrated and defended against outsiders."[8]

At the same time, however, we're not unlike many other people in the world who put their families or traditions or sense of place first. Like many others who are taught to *stay* near family, to *stay* with the ways and means with which you grew up, or to *stay* with your traditions—like insisting that your family *actually* matters most, more than money, more than your career—we do have a tendency to stay through less-than-optimal educational and professional work opportunities at the expense of better ones that lie elsewhere. Loyalty to family, loyalty to a culture (as well as loyalty to the idea—and experience—that we may not fit in elsewhere), and loyalty to the incredibly beautiful, bountiful land carries a strong bond between natives like *us* and *here*. It is here that I—we—make sense.

But it's not like we're unaware of the exploitation and destruction around us or that we see only the beauty of the place. Or that we're unaware that most of the state's bountiful resources are owned by absentee corporations or controlled by other outside interests. In the beginning, as the story goes, when outside interests came into this area to exploit both the state's resources and its citizens, it's not as if no one among us realized what was happening even then. Many of the state's citizens chose to leave under these conditions, of course. Thousands have left and continue to leave, beating roads in all directions out of the state. We're well aware about that fact here: West Virginia's population continues to decline year after year, with many seeking work and better conditions elsewhere. Many of these folks, like Joshua, hope to return one day. But many people—then and now—can't easily just pick up and leave or find better opportunities elsewhere. Many have no money, no resources or skills to offer, and nowhere to go, even if they had the determination to do so. And then there's family, tradition, and land that you must abandon, which can be especially hard. For many of us who can leave but choose to stay, perhaps we can try to heal here, to repair our families, traditions, and land.

## Enter Activism

Many citizens living here cut their teeth on being active in change movements, and although they could leave, they have consistently chosen not to. They prefer to stay and fight. Their resolve is not about stoicism, though. It's about justice.

In addition to having longtime ties to this place, Julian Martin is also one of those activists. He has lived in Charleston for many years and has been vocal through political gatherings and via the news media in fighting mountaintop removal. He lives in a charming cottage near the Kanawha State Forest.

Like many, Julian is angered by what he feels is a corrupt and broken political-industrial system in West Virginia. When the crisis occurred, he knew who was to blame. "I immediately knew," he said. "It was the coal industry. That [MCHM] was a chemical to clean coal. If the coal industry wouldn't be here, that chemical wouldn't be here. . . . It's another one of those hidden costs of coal."

Julian has, for many years, written op-eds for the *Charleston Gazette*, the state's largest newspaper, regarding industry-related environmental issues. He mainly discusses the destructive practice of mountaintop removal. But he also has a personal stake in the issue: mountaintop removal surrounds his family's traditional home place on the Coal River. His "family's history," he pointed out, "is about resisting this stuff." So when news of the chemical spill began to surface, Julian was not surprised; it was "bound to happen eventually," he said. And he made the point in a *Charleston Gazette* op-ed on the water crisis titled "Do They Expect Us to Pay for This?" on the many costs to consumers.

Julian has a clear-eyed conviction of what must be done to improve West Virginian lives and the state's environment. He feels that the water crisis has stirred many to pressure the government to pass laws protecting water resources. "Once you affect your everyday living," he said, "then you get people more involved. They want a clean environment then." As many others we interviewed pointed out, the event mobilized more people to get involved who might not otherwise have been. Speaking about the organizing that happened around pressuring West Virginia legislators to act on the original Aboveground Storage Tank Act (described in the introduction and chapter 1), Julian was quick to point out that "this is the first time I've ever seen community pressure to cause the legislature to pass a decent law" concerning water contamination.

We should point out here that West Virginia is a state where early labor laws were forged from numerous industrial accidents, resulting in injury and deaths, exploitation through child labor, and massive unfair labor practices. The state saw the birth of several laws that altered the working population's lives through helping to establish the eight-hour workday, child labor laws, and safety regulations. Unionism came a little more slowly than in some surrounding states, but it came, eventually, and vigorously, as stories of Mother Jones and the Mine Wars illustrate. Things have changed now, though.[9]

Rallying people toward a just cause can be more difficult in West Virginia than it can be in other places, but the chemical spill was somewhat different because it affected everyone, union or not. Of course, activists like Julian were already active and vocal. But as the event prompted widespread concern, people like Carla McClure who hadn't been involved in activist causes became involved in activism after the spill.

Carla is a thoughtful, intelligent, methodical person who succeeds at everything she's ever tried. Lauded as a writer and editor, she currently works from home for a small business contractor that specializes in public education and health near Washington, D.C. At the time I (Cat) interviewed her, she lived in a small house on the banks of the Coal River with her husband and son. Only a part of Saint Albans was served by the American Water Company, and she had the misfortune to live in that part.

A native West Virginian, Carla is no stranger to industrial disasters and environmental degradation, though she did not live in a part of the state that has undergone continual environmental assault, such as the southern coalfields. She saw firsthand the toll asbestosis took on her father, who was forced into early retirement and died a premature death after working for years in Kanawha Valley's chemical industry. Still, as she was living in a suburban area with relative peace and quiet, the spill caught her off guard. Carla always looks for the good in any situation, though, and one particular good she noticed after the chemical spill was the coming together of strangers as a community so that an incident like this one would never happen again.

Carla doesn't like conflict, she said, so she wasn't primarily drawn to activism. But after the spill, she said, "I started hearing about people who wanted to get together and do something—though at that point nobody was sure yet exactly what, other than let's see what's happening." As time wore on, she was thinking more and more of the younger generation, she said, that she owed it to them to do something. She was especially frustrated that the state's political and community leaders seemed to have little regard for the future generation. "I realized, okay, I'm over fifty now," she continued, "so I'm one of the elders of the community, just like the governor and all his accomplices and the people who are running these companies [involved in the spill]. Whether they realize it or not, they're the elders of our community, or are supposed to be." If the leaders don't seem to have an interest in creating a safe environment for its citizens, she thought, then perhaps it was time for her to push aside her aversion to conflict and act.

So not long after the spill, she decided to attend a meeting at a local church. "I was curious about what people were thinking and feeling" she said, "but at this point people weren't really ready to be activists yet. They just needed to talk about what had happened, and to vent. . . . I mean, I was there to take action, so . . . at some point I had to realize it was okay, that 'these people just need to talk.'"

Eventually, community action became more organized, but because Carla's job is so time consuming, she eventually chose to work with a group that had a presence online. It was, at times, frustrating because her efforts at recruiting others to become more active in the cause were often met with passive responses. But Carla still felt compelled to do what she could. "There was a website that somebody started called 'Easy Action of the Day,'" she reported, "and I had been doing that. So out of that [initial] meeting I committed to continuing to do those easy actions. Like maybe it was calling somebody or emailing an office about a bill that was coming up or that kind of thing. And so we did that, you know, wrote a lot of letters, talked to people."

In that work, she found others who also wanted to act. Still, though, many fell back into a familiar attitude, saying "well, this is life, and this is society, you know, and this is the way things work right now, or don't work right now. There are lots of chemicals out there. This is only one hazard." Activism in any form, is never easy, but thinking about the future kept Carla "thinking about the young people now [and how] they don't realize that this could get a lot worse before it gets better."

## Notes

1.  All quotes by Anita Edmonds in this chapter are from Anita Edmonds, interview with Cat Pleska, August 17, 2014.
2.  All quotes by Belinda Vance in this chapter are from Belinda Vance, interview with Cat Pleska, July 7, 2014.

3. All quotes by Paula Clendenin in this chapter are from Paula Clendenin, interview with Cat Pleska, July 7, 2014.
4. Sue Brookshire, interview with Marla Griffith, July 23, 2014.
5. All quotes by Matthew Chesebrough in this chapter are from Matthew Chesebrough, interview with Cat Pleska, July 11, 2014.
6. All quotes by Carla McClure in this chapter are from Carla Thomas McClure, interview with Cat Pleska, July 10, 2014.
7. All quotes by Julian Martin in this chapter are from Julian Martin, interview with Cat Pleska, July 9, 2014.
8. See Jay Thomas's discussion of Waggoner's essay in chapter 5. For a recording of Eric Waggoner reading his entire essay, see Scott Finn, "'To Hell with You'—A West Virginian's Raw Response to Water Crisis Goes Viral," West Virginia Public Broadcasting, January 18, 2014, accessed September 24, 2017, http://wvpublic .org/post/hell-you-west-virginians-raw-response-water-crisis-goes-viral.
9. For example, in 2016 the West Virginia Senate and House of Delegates passed right-to-work legislation. Then-governor Tomblin vetoed the measure, but the legislature overrode the veto and the new law went into effect on July 1. In August a temporary injunction went into effect, delaying the law's implementation. The 2017 legislature passed legislation meant to address the issues that allowed the injunction to go forward; current governor Jim Justice vetoed that bill, saying the courts need to decide. The State Supreme Court, in turn, knocked down the injunction in September 2017. See, e.g., Phil Kabler, "WV Supreme Court Overturns Ruling That Blocks Right-to-Work Law," *Charleston Gazette-Mail*, September 15, 2017.

# CHAPTER 7

# In and Out of Appalachia

Emily Mayes

When I wrote an early version of this chapter as part of a 2016 graduate seminar, I was a student in Marshall University's graduate humanities program. In January of 2014, though, I was a recent college graduate living and working in the metropolitan D.C. area. I spent my days in the fast-paced city, paying my dues (and parking tickets) and looking forward to what I was sure, in five to ten years, would be an extraordinarily bright future. My past was in West Virginia, but now I was in the nation's capital, in the heart of the city, and nothing was more intoxicating than the pulsing energy that surrounded me at every turn. I relished the sounds of car horns and clinking glasses and most of all the feeling of life unfolding around me, vivacious and colorful. Truly, at that time, no place could have been further from my mind than Charleston, West Virginia.

On January 10 I had spent a long day at the office, shuffling papers and making coffee, and I was looking forward to an evening out with friends. A metro and a cab ride later, we were sitting in a shabby dive downtown, sipping our drinks and discussing the day, when news of the Charleston water crisis flashed on the television screen just across the room. I listened carefully to the perfectly coiffed newscaster and was suddenly jolted with concern for my family and many questions about the chemical that had somehow managed to poison the water of some three hundred thousand residents. What I remember most, though, were the reactions of those around me.

Almost immediately, there was a simultaneous myriad of heavy sighs, eye rolls, and blustery comments about the need for hillbillies to acquire a little common sense. One man in a plaid peacoat sitting just a table over from us exclaimed: "Those damn idiots! Whaddya think was going to happen living around a bunch of damn chemical plants? They ought to pick up and move, every one of them!" But for most of the people in the room, the news hardly struck a chord. Many could not even be bothered to look up from their half-empty glasses because there, in the center of a thriving metropolis, it hardly mattered what was happening to faceless people in a place as far removed as southern West Virginia.

Still, as enchanted with my current surroundings as I was, what happened in Charleston mattered to me. My family was there, as were my memories, and I couldn't imagine that in a time of crisis, people could be so callous and unconcerned. In those moments, it was readily apparent that even if I cared about the people of Charleston, even if it was a place big in my heart though small on a map, no one else much cared. Not outside of Appalachia. So I spoke with my family, and as days without water turned to weeks, I sent money and hoped for the best. But I have never forgotten how the water crisis was received in Washington. Everyone seemed content to shrug it off, to think of it as an isolated incident involving people who should have known better than to reside in a place called Chemical Valley. The general consensus was that it had happened to "them," not "us," leaving me to feel divided and unhinged. My life in Washington and my memories in West Virginia couldn't quite successfully coexist, it seemed . . . I was deeply uncomfortable.

The truth is that West Virginia has been known for its struggles for years, and it continues to struggle today for various social and economic reasons. The result is a widespread belief that West Virginians are too ignorant and too unmotivated to cultivate progress for their state, even after a disaster as severe as the 2014 water crisis. Though the state may not look like much to some, it is home to me. And whatever challenges exist within the state boundaries, they are authentic to the land and the people. They are the legacy of a past that continues to define the present. The water crisis isn't an isolated incident, but it is one particular event that illuminates realities of the Appalachian experience, realities not often considered by critical outsiders.

After the water crisis, my life changed dramatically. I left D.C. and relocated to Charleston for two years while completing my MA in the humanities. During those years, while reconnecting with family, local water became more central to my life in ways I never expected. Not only did I wash clothes and bathe in it; I fished in the rivers and, if this chapter is any indication, spent a great deal of time thinking about how water can impact lives. Each time I took a drink of water in Charleston, like many residents, I couldn't help but think that perhaps the water wasn't entirely safe, and each time I sat on the bank of the Kanawha River casting my line out, I imagined that I must be exposing myself to all kinds of potentially dangerous chemicals. Still, Charleston was and is my home, the land of my family. It's a part of who I am. So I took the risk.

Though as of this book's publication I have graduated from Marshall, have left Charleston again, and am now living and working in North Carolina, the water crisis, I think, was ironically positive in that it led to a stronger sense of community in Charleston and encouraged individuals to think more critically about their community, their state, and their water. Yet from my view back in Washington, it also widened the divide between Appalachians and non-Appalachians, inspiring "us and them" rhetoric that reinforced a dominant image of West Virginians as lazy and idle.

In truth, there is no denying that West Virginia has its problems. I only hope that outsiders will come to see that those problems are a direct result of other, complicated factors, such as industrialization, that are beyond the average citizen's control. The history of coal and other industries helped to augment widespread poverty as these industries came and went to extract precious resources. These historical circumstances also affected education levels, making for a tradition of complacency that is continuously corroborated by a kind of "love it or leave it" Appalachian identity. It is a vicious cycle that proves, even now, difficult to break. But West Virginians, in spite of their struggles, do matter, and with incidents like the water crisis serving as catalysts for change, perhaps the state can overcome obstacles to progress and move forward toward a brighter future. Maybe.

There is no denying that the water crisis has led to an increase in community altruism and political participation in the area, which is the subject of previous chapters. Some may even argue that it is has proven to be a real catalyst for change in West Virginia. As the water crisis oral histories illustrated, during and immediately following the crisis, residents assisted their neighbors, offered fresh water to those in need, and checked in on the elderly and disabled. Since then, they have begun organizing committees, questioning the health standards of the local water company, and pushing for progressive legislation. Residents here recognize that their involvement in activist efforts may be the only way to force government and corporations to make social welfare a priority. Activist and lawyer Paul Sheridan's words stay with me: "It's sort of an awakening and, you know, maybe sort of like the crisis. You say, oh the water doesn't automatically come to our taps wholesome, pure, and safe. There's a process involved and things can go wrong, and so what is it and how can we help, or can it be made better?"[1]

Still, here in West Virginia change often doesn't come easy. Generating common, active interest in the safety of water for consumption and successfully implementing new policies to address that issue can prove especially difficult in this area. Resources are scarce, and there exist myriad complex social issues that center on the industrial mainstay here in Appalachia, namely coal. West Virginia is home to unique values and concerns that in many ways counter the outcry for social and political change that came along with the chemical spill. And so many say that calls for safer water, more competent and empathetic government officials, quicker response times in the case of emergency, and better means of detecting unsafe conditions in public water supplies may never be answered. But as the oral histories attest, many others refuse to believe or accept this.

Perhaps one of the greatest challenges to change or progress in West Virginia is the state's long-standing relationship with the coal industry. Though the coal industry employs only a fraction of the state's population, it still remains at the center of the state's politics, culture, and identity. For many of those who work in coal and its related industries, coal is not just their livelihood; it is a way of life, a steady pulse that continues to keep this state alive. And especially for those people, considering the state's, or one's own, identity outside coal is nearly impossible. "I think a lot of people," said Dave Mistich (interviewed by Jim Hatfield in chapter 9), "are really quick to say the coal mines have done so well for my family, or this industry has done so well for my family, I can't say a bad word about it."[2] For those who depend on the successful operation of the coal mines and chemical plants to support their families and retain their independence, even when those same industries may lead to environmental degradation, many West Virginians wholeheartedly continue to put their faith in coal and organize their lives around it.

Predictably, West Virginia's obsession with coal can put a serious strain on any attempts at activism or social reform brought on by the water crisis. Though I doubt anyone would argue against the merits of safe, clean water, given the hard choice between their health and their livelihood, many West Virginians would have to put reservations about their safety aside in favor of their income. And coal, for some, is a primary source of income. In terms of West Virginia state government, this same commitment to coal further creates challenges. Whatever health and safety concerns coal and associated chemicals pose, many West Virginia officials are well aware that their successes are tied to the commercial successes of coal, so they will continue to defend it, even at the cost of public health.

Another related challenge to change or progress here is the reality of poverty in this state. While many West Virginians are quick to focus on more "positive images" of West Virginia, fact is, poverty remains one of the foremost issues in West Virginia, year in and year out. According to the national census, between the years of 2009 and 2013, 17.9 percent of West Virginians were living below the poverty level, compared to 15.4 percent of Americans.[3] A high poverty rate often directly correlates to lower levels of education, for example, and as the West Virginia Department of Education reports, most of the schools in the state perform at a level under par. As a matter of fact, according to their research, of the 650 public schools in West Virginia, 341 perform at a critically low level. Of these, 29 are categorized as priority institutions, meaning that students' performance in mathematics and reading are the lowest in the state, while 97 are considered focus institutions, meaning that there are disproportionate achievement gaps within the student population. The remaining 215 are considered support institutions, which indicates that these schools have not met targeted testing scores and that most students have not achieved satisfactory academic progress.[4]

A high poverty rate coupled with low academic achievement creates real problems for West Virginian residents. It perpetuates a gap between the needs of the people and the opportunities available for them to have those needs met. Poverty that begets a low-level education leaves students, in many cases, ill equipped to elevate themselves above the poverty in which they live. It lowers expectations of achievement and narrows priorities. Though many often neglect to mention it, this is part of a larger set of obstacles Charleston faces as the residents attempt to restore the community's confidence in their water and their government.

The roots of these problems run deep and are complicated, of course, but West Virginia's unsettled past is at least partly to blame. Unlike other states in our country, West Virginia has a long history of absentee ownership of land and resources, in which people who don't live here make major economic and other decisions for those who do—as they extract the state's riches and spend and invest the profits elsewhere.[5] (In Wyoming County, for example, as much as 90 percent of the county's land base is owned by corporations not based in the state. The situation is similar in other counties.[6]) This experience of living under absentee ownership in West Virginia goes back generations to before West Virginia was even a state, a situation that many coal and other extraction companies used to exploit workers and landowners in ways that are well known and well documented. It was common, for example, for coal companies to require miners who were already underpaid to lease their mining tools or for the coal company to enlist a number of procedures, such as cribbing (a practice that enlarged the size of coal cars but paid workers the same for the car's load), that were intended to boost production while maintaining low wages for miners.[7] Since long before the state's industrialization, through coal mining booms and busts, West Virginians have been subject to continued detrimental treatment, considered marginal and unimportant working parts of an industrial machine based beyond and outside of the state.

One might say that times have changed today, but it seems to me that the consequences and legacy of West Virginia's peculiar past still survive. Placed into its present context, it is hardly surprising that for many residents, concerns about water safety, human rights, and the possibility for positive change are relatively minor in comparison to the demands of day-to-day living. Many are consumed with financial anxieties, and many are still locked in poverty and armed with a deficient education that is perpetuated by that poverty. Not to mention that, for some, it is likely that the necessary steps and protocol for working to produce tangible results in the way of progress are not readily apparent. And so the issue of poverty stands strong against improvement in West Virginia, as do deeply ingrained traditions.

Along with the legacy of coal, ensuing poverty, and the lack of education is another tradition here in West Virginia: complacency. There, I said it. There are lots of wonderful traditions here, like family, old-time music and dance, good neighborliness, and food. But if you live here, you know that this is a tradition, too. And I know I should be very careful about saying this, especially as it brings up victim-blaming theories of poverty. Be that as it may, trapped in the conventions of the state's poverty and absentee ownership, it should not be surprising why some here readily accept unfair and or unsavory circumstances—they often expect nothing else. Spills and contamination do happen in places like Chemical Valley and elsewhere in the state, and while there are many ready to stand for justice, change, and action, many others just don't have the resources or knowledge available to stand firm against social and environmental injustices.

Several of the stories that this book's oral history team collected spoke to this issue. Matthew Chesebrough (introduced in chapter 6), is, like me and so many other young(er) West Virginians, skeptical about the possibility of real and inclusive change for the state, given its past and still unfolding legacy. From his point of view, West Virginians have been mistreated for years—so often, in fact, that it's become not just a tradition but a *dominant* tradition, deeply engrained in the Appalachian consciousness. He doesn't believe that improving water systems and strengthening government regulations is a realistic goal when positive change has traditionally been so hard fought in the Mountain State and with so few people standing ready to combat social injustice. "Nothing has changed," he said.[8] I admit that it's hard for me to disagree.

Related to this issue of tradition is how it plays into West Virginian identity. Questions of identity are complex, and people draw differently from different wells of experience to make their lives meaningful and then project an image of self to others and themselves. But, again, if you live here, there's no denying that part of the West Virginia identity (of both the state and its people) is built on the tradition of social inequity, then shaped by outside perceptions of what it means to be an Appalachian. In regard to tradition, one of the resulting Appalachian identifiers is a sense of pride in managing to take on difficult circumstances and survive, whatever the odds. West Virginia has endured hardship time and time again, having persisted in spite of forced industrialization, economic downturn, outward migration, and even drug infestation. The state's ability to survive all of these, as many residents see it, is a testament to the people's resolve and tenacity, and for them, the water crisis is nothing more than yet another obstacle they have successfully tackled, evidence of the tenacity and resilience of which they remain exceedingly proud. When I read Eric Waggoner's essay on the Elk River spill (discussed in chapter 5), especially where he wrote, "To hell with all of my fellow West Virginians who bought so deeply into the idea of avoidable personal risk and constant sacrifice as an honorable condition under which to live, that they turned that condition into a culture of perverted, twisted pride and self-righteousness, to be celebrated and defended against outsiders"—it struck a chord here, and not one faintly heard.[9]

Outside perceptions of West Virginians, however, are a different matter. West Virginians like me know that many negative perceptions of the state and its people thrive outside of Appalachia. We often like to think that it doesn't affect us, but it does. Arguably, outside perceptions color some West Virginians' personal sense of identity. Sometimes we ask ourselves: If no one outside of Appalachia cares about what happens in West Virginia, and if outsiders are convinced that we, by virtue of our Appalachian heritage, are necessarily indolent and ignorant, then how can West Virginians think any differently about themselves? Can West Virginians really strengthen their communities and work toward positive change with no support from the outside?

Perhaps. But it seems that with the state's absentee ownership, coal legacy, rampant poverty, educational levels, and all things related, that this complacency may also come from an ingrained and degraded sense of self-worth, that we don't really matter to the country as a whole. Sitting there in that dive in downtown D.C. back in January 2014, it seemed that way. And as the crisis unfolded, it also seemed to me that "not mattering" and a degraded self-worth was, too, as much a part of West Virginia identity (cultural or otherwise) as that strong-willed fortitude that everyone likes to bring up. If that's truly the case, then that poses a real challenge to progress of any kind in the state—concerning a water crisis or any other issue. From where I'm sitting, progress seems like an insurmountable task.

# Notes

1.  Paul Sheridan, interview with Jim Hatfield, August 26, 2014.
2.  Dave Mistich, interview with Jim Hatfield, September 13, 2014.
3.  QuickFacts: West Virginia, United States, United States Census Bureau, accessed September 4, 2019, https://www.census.gov/quickfacts/fact/table/WV,US /IPE120217.
4.  West Virginia Department of Education, accessed July 31, 2016, database searchable at http://zoomwv.k12.wv.us/Dashboard/portalHome.jsp.
5.  On the long history of absentee ownership in West Virginia, see, e.g., John Alexander Williams, *West Virginia: A History*, 2nd ed. (Morgantown: West Virginia University Press, 2001).
6.  See Samuel R. Cook, *Monacans and Miners: Native American and Coal Mining Communities in Appalachia* (Lincoln: University of Nebraska Press, 2000), 278.
7.  See Christopher Dorsey, *Southern West Virginia and the Struggle for* Modernity (London: McFarland, 2011), 42–52.
8.  Matthew Chesebrough, interview with Cat Pleska, July 11, 2014.
9.  Jay Thomas discusses Waggoner's essay in chapter 5. For a recording of Eric Waggoner reading his entire essay, see Scott Finn, "'To Hell with You'—A West Virginian's Raw Response to Water Crisis Goes Viral," West Virginia Public Broadcasting, January 18, 2014, accessed September 24, 2017, http://wvpublic .org/post/hell-you-west-virginians-raw-response-water-crisis-goes-viral.

# Exploring the (Human) Nature of Disaster: Impact and Responses

Brian A. Hoey

Beyond laying out particulars of setting and our approach in this project, the introduction provided our first detailed account—by way of extended transcript excerpts—of the lived experience of conducting and participating in an oral history interview within the context of this collaborative effort. Our insight as readers extends from the kinds of questions asked to how conversations progress narratively to reveal something of the *process* that those affected by the chemical release underwent, from first learning about the unfolding crisis to practical and emotional reactions, as well as an ongoing effort to make sense of its physical and psychological impact. In the latter, the reader confronts the profound discomfort shared by those affected that impacts their quality of life and, for some, precipitates difficult personal decisions about whether to stay or leave the state—the central theme for part II of the book. In concluding part I, Trish Hatfield's chapter brought us further into contact with varied dimensions of experience—beginning with ways those affected came to realize that they were at risk. From her own background of moving to the Kanawha Valley from out of state to settle down with her chemist husband, Jim (from whom we hear in part III), we learn of opportunities afforded as well as challenges presented by the many chemical companies that dominated the region—including how her family simultaneously developed escape plans from the Chemical Valley while coming to accept the risk that they faced daily by staying despite demonstrable dangers. We are also given some insight into decisions and practices that led to a dangerously centralized water system under the control of a single company, the West Virginia subsidiary of American Water. Building on Beth Campbell's interview with Rebecca Roth in the introduction, Hatfield provides us with a deeper sense of how oral histories were collected, as well as the evolving role of those engaged in documenting the experiences of others, while translating her own encounters into the personal accounts that are the heart of this book.

Opening part II, Jay Thomas's chapter furthers our appreciation for the process by which those affected would come to learn of the chemical release, which put us in touch with how, as a local businessman, he navigates a shifting relationship to place. Owner and co-owner of two restaurants in Charleston— at the geographic and symbolic center of the affected area—Thomas learned of it while traveling out of the state. His story relates how, at a distance, he was able to pick up details of what happened through conversations with others who were hearing of the leaking Freedom Industries storage tank in news reports. He returned to Charleston with the need to keep his businesses afloat while also having been changed as a man who now carries feelings of betrayal, loss of trust, and an abiding embarrassment for what happened and how others perceive his community. In chapter 6, we heard from native West Virginians raised in rural parts of the state. At the time of the release, Cat Pleska and Joshua Mills both lived outside the affected area and thus maintained some practical distance in day-to-day routines, yet they relate how their experiences growing up in the state have always kept them in the shadow of economically and politically powerful industries that shape their sense of self and place in a multitude of ways. Their chapter conveys important distinctions between the experiences of those affected engendered by their rural or urban residential status. Picking up on a thread from Jay Thomas's chapter, Pleska and Mills further a conversation had by many in the weeks and months following the chemical contamination of their drinking water about whether to stay or leave the state, given the depth of its multidimensional impact on personal relationships to place.

Finally, Emily Mayes—also a West Virginia native but one who came to learn of the release while living outside of the state—reflects on its meaning at a practical and somewhat emotional distance. Residing in Washington, D.C., at the time, Mayes offers how, upon hearing news of the contamination with others in a crowded bar, she was given a powerful reminder of how many people outside of the state view those within with either indifference or a disgust that manifests in collective head-shaking at the apparent haplessness of residents who—it is reckoned—should not be surprised to find themselves with contaminated drinking water in a place that they willfully accept as Chemical Valley. Dispiriting reactions led Mayes to experience and then describe in these pages her complicated feelings about West Virginia as a place where complacency, as she describes it, has become a tradition. This leaves her with deep skepticism about possibilities for transition to a future beyond what seems an overwhelming dependence on powerful industries whose activities carry great risks to public health.

# Impacts

As I suggest in part I, many social scientists emphasize the practical importance of distinguishing between what might be characterized as "expert" and "lay" forms of knowledge, especially as related to risk—specifically, how such knowledge is constructed and the meanings those constructions convey. While those involved in this project resist simple characterizations implied in making this distinction, for my purposes, it is important to carry on with the distinction given how fundamental it is, not only within the scientific literature generally but also in practical ways, that "sides" are formed (or at least perceived) in the seemingly inevitable public confrontations that follow disaster. For the most part, the distinction I refer to is a socially meaningful one made between those who hold recognized (generally scientific but also possibly legal) credentials and would thus be considered "expert" and those who do not—regardless of what practical experience they may have—and are considered nonexpert or "lay" as a result. Further, in most confrontations over feared contamination by toxins, these experts are typically nonlocal. To some extent, this distinction is blurred in the Kanawha Valley, where many citizens, because of their work in the chemical industry, hold advanced degrees in chemistry and engineering. As we hear from Jim Hatfield in chapter 8, at least some of these credentialed locals were actively engaged in the months of confusion after the chemical release, writing op-eds critical of both government and industry as well as participating in public campaigns to support postcrisis investigations that would open routes to safer public water systems.

Unlike the generalizable knowledge of experts, lay claims to knowledge are shaped in the context of a particular social and economic context and a person's own networks of communication and are based on people's everyday experience.[1] The discontinuity between expert and lay perceptions of events and specifically the tendency within scientific ways of knowing to discount the role that everyday experiences play in shaping the psychosocial and physical health effects of disasters—together with the social crisis situations they entail—is an essential starting point for my discussion of how disasters like the West Virginia water crisis may be experienced as well as how experiences of disaster shape individual and collective responses over the short and long terms.

The question of *time*—specifically how we might grasp the time frame of a disaster—is in many ways at the root of considerable misapprehension. Drawing on his anthropological research on the impact of hurricanes, Ben

McMahan explains that although both media coverage and federal disaster recovery efforts treat disaster as an acute event with discrete temporal and spatial boundaries, affected communities are complex sociocultural contexts that include both long histories of comparable experiences and persistent effects from any particular event long after the media and national attention have moved on to other, more immediate subjects.[2]

Within the disaster literature, there is much conversation on the extent to which researchers may find psychiatric disorders such as post-traumatic stress disorder among affected persons. While emergent public health attention to the potential for long-term psychological impact and the possible need for continuing treatment suggested by this diagnosis may be taken as a positive development, Doug Henry suggests that as a product of Western biomedical categories and with a therapeutic regimen that is "shaped by Western ideas of cognitivism, in which trauma is located as an event inside a person's head, rather than representing a *social* phenomenon," even diagnoses such as post-traumatic stress disorder may limit expert perception of the problem's full set of causes.[3] This limitation may then preclude a holistic, systemic understanding in which, for example, individual recovery may be bound inseparably to both physical and social recovery at the level of the community as a whole.[4]

Although it is undeniably true that *individuals* experience psychosocial stress, this experience is shaped by the particular sociocultural and physical context, which may be responsible for either increasing or decreasing its harmful effects—just as this context is responsible, at least in part, for the relative vulnerability to disaster of different populations of community members in the first place. Again, disasters are highly *meaningful* events. As such, individual and collective responses to the experience of disaster are not simply reactions to physical effects. Rather, these responses arise out of the complex combination of physical effects acting together with how events are interpreted by those affected—what they come to represent for different people—in what are ultimately *socially* constructed and experienced crises.[5]

Given this fact, such differentiating factors among members of a community and between distinct communities as socially defined race, ethnicity, class, age, and gender not only give shape to a particular "riskscape" for different populations but also shape patterns of ideological consensus and conflict in the wake of a disaster.[6] It is common to find pre-disaster systems of social relationships responsible for structuring inequalities, for example, that serve not only to exacerbate preexisting suffering of certain groups—for example, as may be related to their limited access to essential resources—but also to add to their distress by justifying prejudicial treatment in any official response.

In his research on the question of how prevailing moral codes may structure the provision of relief following famine in India, William Torry demonstrated how religiously sanctioned inequalities within affected areas shaped social adjustments during crisis in ways that were not "radical, abnormal breaks with customary behavior; rather they extend[ed] ordinary conventions."[7]

Although one of the most powerfully revealing narratives to emerge from disaster research is the fact that disasters are, despite common expectation, all too often predictable continuances of existing conditions, recognizing this continuity in no way precludes our appreciation for what are also substantial breaks or, perhaps, *dislocations* between pre- and post-incident conditions. In my earlier chapter, I discussed how preexisting conditions can influence both the vulnerability of particular populations to disastrous episodes as well as post-disaster responses. A good deal of attention has been given in the academic literature to how what is commonly referred to as "social capital" may affect the capacity of individuals and groups to *respond* to the experience of disaster—something that may be categorized by relative degrees of what may be called "resilience." Both ideas are worth closer examination. On the one hand, social capital in this context may be thought of as a measure of the resources available to either individuals or groups as determined by the nature and extent of their own social networks of obligation and membership in a given community.[8] On the other hand, resilience as a concept has tended to be invoked in order to provide some measure of what may be called "societal resistance" to hazards or, specifically, the capacity to withstand at least the most potentially harmful effects of exposure to either hazards or the experience of living through disasters themselves.[9]

Resilience has typically been used in the context of ecological science. In at least one effective illustration, we can see how social scientists make an ecologically informed application of the concept of resilience. In this case, it is relevant to show how resilience can be treated as a "capacity of linked social-ecological systems to absorb recurrent disturbances such as hurricanes or floods so as to retain essential structures, processes, and feedbacks," which is to say that the "systems" in question are capable of *self*-reorganization in the wake of disastrous events as opposed to depending wholly on outside intervention.[10] The term *adaptive flux* has been used to described such self-organizing, long-term coping strategies used by local peoples that anthropological and other research has shown can enable them to persist and survive under what might be variously harsh or uncertain conditions and what are—as is the case with many indigenous populations—socioeconomically and geographically marginal locations.[11]

Anthropologists, in particular, have documented over the past century how small-scale, local, or "traditional" societies have suffered diminished capacity in terms of their hazard management and resilience as a result of integration with the global economy—especially in the context of large-scale, economic interventions such as natural resource extraction and energy production projects including hydroelectric dams. There is no reason to believe that diminished capacity or resilience in such contexts is an experience restricted to traditional societies; in other words, we could likely apply the same analytical approach to many places like West Virginia, given the long history of such interventions here.

Not surprisingly, quite a number of factors—including social capital—are thought to affect an individual's or community's capacity for resiliency in times of crisis. Social capital in particular has been the subject of much study within the context of social crises such as disaster. Typically, a reserve of social capital is widely considered a kind of protective buffer against potentially damaging levels of environmentally (both physical and societal) induced stress. Studies such as Francis Adeola and Steven Picou's, which assessed long-term patterns of psychosocial distress following Hurricane Katrina, suggest that, unfortunately, it is also the case that tragic events such as Katrina can *undermine* or weaken the positive, protective role of social capital as people are displaced and sometimes permanently relocated following disaster.[12] All of this can lead to the loss of critical social relationships and even to the total collapse of personal networks that had supported individual social capital.

We may also see social capital—through the very same, though intact, personal networks of obligation—potentially *exacerbating* psychosocial impacts where, under certain conditions, a burden of reciprocity (or, in this context, the need to provide aid to others with whom one has an established relationship) weighs excessively on an individual or household. One might say this is part of larger potential risk associated with degrees of what we could call "attachment." For example, strong attachment to a given social community and to a specific, geographic place that holds dense personal and cultural meanings, has been shown to create more profound stress in the wake of disaster.[13] As introduced in my earlier chapter and further revealed in part II, we can see that place and place attachment profoundly shaped the impact of the West Virginia water crisis. Speaking on attachment to place, environmental psychologist Marc Fried notes that "place attachments which become intense and exclusive can preclude all alternatives or even information about alternative future potentials. Thus, they negate the dialectic interplay between changing needs, desires, conditions, and overt transitions. These become dysfunctional when adaptation to new growth opportunities or possibilities of greater gratification in personal or community life are renounced."[14]

Acknowledging that place attachment, generally connected to such constructs as belonging and considered individually and collectively healthy, can lead to negative outcomes sheds light on how the concept of "corrosive community" is used in disaster research to describe where technological disasters can lead to long-term individual and communal harm, beyond anything that the disaster itself may have caused directly. These include the indefinite nature of harm from toxic contamination, as well as a lack of access or control over information, the effect of litigious discourse dominated by lawyers who seek to minimize liability for corporate clients while often blaming victims, and the disruption to basic ways of seeing the world that may be caused by loss of trust in basic social institutions and service providers due to what has been generally referred to in the literature as "recreancy."[15] Specifically, recreancy refers to when individual or institutional actors fail to carry out socially or legally defined responsibilities and, as a consequence, threaten societal trust.

As Krista Harper explains, living with both the certainties and uncertainties of environmental risk, which her work explores in the context of the Chernobyl disaster, exposes not only the limits of the state's ability (or its will) to protect its citizens but also the limits of citizens themselves to protect their homes and families.[16] Anna Willow captures this latter sense of losing control or disempowerment in her work with parents who, while living near hydraulic fracking sites, express strong feelings that governmental representatives have effectively taken away their ability to keep their children safe.[17] Others, such as Stephen Kroll-Smith and Stephen Couch, note how within what Michael Edelstein refers to broadly as "contaminated communities," social dysfunction may manifest in the form of growing anomie among community members.[18] Anomie as a concept may apply to both the individual and to larger collectives—even society as a whole. At the individual level, it incorporates a sense of personal unrest or what may be called "alienation," coupled with an anxiety thought to arise from the perceived lack of control over events. It may also be applied to society and in particular to indicate social instability that emerges from a breakdown in the standards and values that are taken as central to maintaining social order.

As Erikson saw in the case of Buffalo Creek, for disasters deemed human-caused by those affected, a responsible party (or parties) can and will be blamed.[19] As Duane Gill and his coauthors noted, while this collective desire to hold someone accountable for mass suffering may be expected, the very fact that placing blame is possible in the context of technological disasters can serve to "heighten anger, frustration, fear, hostility, loss of trust, and uncertainty" all of which then "contribute to prolonged psychological stress" and possible anomie.[20] William Freudenburg noted that, unlike what has been observed following natural disasters—where a healthy phenomenon generally referred to as "therapeutic community" emerges out of devastation as people work together to rebuild physical damage to their shared environment—in the aftermath of technological and, ultimately, *human* failures such as the Buffalo Creek flood, we may see a flipside phenomenon. This is what Freudenburg described as the aforementioned *corrosive* community.[21] Corrosive community is pathogenic.

In their application of this concept, Steven Picou and his coauthors have suggested that although researchers have identified many contributing factors, in their own work they have found none of them as harmful as the litigation process.[22] This is despite the fact that litigation is often intended by those who seek to formally assign culpability as a means of achieving more than simply a kind of restitution for damages. Although it may not often be achieved, injured persons may attempt to use the law to pursue a *therapeutic* outcome. This may be referred to as "closure" in the context of recovery from traumatic life experiences. Instead, litigation does anything but provide a conclusion, as it only further unsettles—often exposing "experts, or specialized organizations, as irresponsible, incompetent, and untrustworthy [and thus] contribut[ing] to a persistence of chronic disaster impacts through loss of trust in traditional institutional support systems."[23]

Relatedly, in a sensitively documented account of the lived experience of residents of Endicott in upstate New York, Peter Little examines the role of mitigation efforts in shaping that experience.[24] The community is affected by a toxic underground plume of chemicals that have remained long after closure of an IBM industrial facility that once employed many in the town. Little points out that vapor intrusion mitigation by way of continuously operating exhaust fans installed in people's homes in order to clear basements and crawlspaces of possible airborne toxins leached out of the ground is fundamentally a "rational-technical" approach to uncertainty, intended to lessen possible negative health impacts. Although both IBM and New York state public health officers envisioned the remedy as a way to provide noticeable reassurance of this ongoing therapeutic effort, Little uncovered a shared mitigation experience of Endicott residents that is both emotionally and socially constituted and deviates significantly and pathogenically from what officials intended.[25]

Though unlike the experience of Endicott residents, given that flushing pipes, ostensibly to clear them of contamination, by those affected West

Virginia residents was temporally limited to each household's own experience of the officially sanctioned process of pipe flushing, mitigation efforts in West Virginia appear to have similarly compounded suffering, both physical and psychical, to lasting effect. Project participants such as Carla McClure describe how flushing created acute distress leading both her and her husband, Phillip, to become faint and short of breath. After checking how her flushed pipes might provide water that wasn't subjectively contaminated, as suggested by smell or taste, Martha Ballman found herself spitting out a mouthful of water that she described as "burning" as if hot.[26] This sensation lasted nearly two weeks. When she later developed kidney disease, she could not help herself but wonder if there might be a connection. Just as litigation may be meant to bring beneficial remediation to damaged places and hasten the recovery of suffering people yet instead may only deepen psychosocial wounds, mitigation efforts may at best be ambiguous in their overall impact.

Studies such as Little's apply social theories of an emotional, embodied subject and phenomenological approaches to behavior that underscore how persons are oriented in worlds of sensory experience—they have relationships to place. These studies also make use of what is known as "practice theory" with an emphasis on a contextually positioned subject and a tension between individual agency and social structure. Anthropologists have used the concept of "embodiment," for example, to draw attention to qualities of lived experience and in particular to how the effects of disaster-related trauma are manifest and expressed through individuals' efforts to understand and react to their physical and emotional trauma in culturally informed ways.[27] These expressions reflect what Gregg Mitman and coauthors refer to as "tangles of economy and flesh" found at the meeting between a physical and social environment and the health of real, flesh-and-blood people in those environments whose experience cannot be either adequately conveyed or understood through statistical data.[28]

As expressed in culturally defined idioms, which is to say that they are distinct to particular cultural contexts, narratives of illness provide a coherent structure or framework for personal experience and identity. These narratives help to capture a particular form of misfortune and, in the context of disaster, what has been called "environmental suffering" and are told in what medical anthropologists refer to specifically as a *language of distress*.[29] While often expressed verbally, illness narratives interpret causality and ultimately help to integrate personal experience into culturally meaningful life histories that may help those traumatized by toxic contamination or serious disease make sense of their experience. Distress can also be conveyed nonverbally, which is to say, through the body itself.

This nonverbal communication has been called "somatization."[30] For example, in their research following a deadly 1985 landslide in Puerto Rico, Peter Guarnaccia and his coauthors examine the lived experience of this disaster while accounting for symptoms exhibited by those affected according to those that could be taken to characterize a "standard" medical condition (i.e., its symptomatology), and prevailing "folk" or even what may sometimes be referred to as "lay" (i.e., culturally specific) illness categories.[31] Their study revealed that popular local illness categories such as *ataque de nervios* (nervous attacks) played an essential role as a culturally recognized complex of symptoms, behaviors, and both verbal and nonverbal expressions that organize how survivors both experience and respond, in this case, to the landslide.

Highlighting the sensual qualities of embodiment, Josh Reno describes how strong odors associated with a southeastern Michigan landfill shape the embodied experience of place for nearby residents.[32] Specifically, these odors provide what he refers to as shared orientation to a perceptual referent, "semiotic indices" of a common experience of exposure to noxious substances repeatedly registered by people as a pathogenic bodily invasion akin to bacteria or viruses. Similarly, in their study of the impact of hydraulic fracking, Anna Willow and Sara Wylie speak to how residents in areas with active gas wells have real concerns about this unconventional energy development that are, from the very beginning, manifest in their bodily, *sensory* experience of their changing surroundings: "Foul odors and discolored waters reveal the presence of potentially hazardous wastes, spills, and/or pollution. Our senses tell us when something is wrong. . . . The useful term *dysplacement* . . . underscore[s] how perceptible pollution can transform formerly positive sensory experiences of place to experiences of profound alienation. In other words, even when people are not physically displaced, the sensory experience of environmental degradation can lead to equally damaging dysplacement."[33]

Such research effectively bears witness to incidents of what Little refers to as *sociospatial contamination* that may entail a "reconfiguration of emotional attachment to place, and the birth of new 'idioms of distress' that are in part conditioned by concern over toxic exposure and frustration with corporate and state decisions and technologies."[34] Within the literature that addresses the psychosocial effects of disaster for people as well as how responsible parties and government officials respond, there are numerous references to meaningful shifts in the relationship between people and place, broadly defined. As I have described elsewhere, disasters—including perhaps especially those entailing toxic exposure and contamination such as in the water crisis—may fundamentally change the nature of this relationship such that a person's immediate environment and what may have once been a source of security and even a kind of steadying anchor for personal identity becomes, instead, a threatening presence.[35] Within this shifting sense of place, the social context may lose its meaning and its role in providing comfort and stability.[36]

While for many, this shift may constitute an abrupt change following substantial disruption in the familiar, reassuring routines of day-to-day lives, for others there is no watershed moment after which they find conditions have changed so significantly that they can no longer have the same relationship with these places that they had before the disaster.[37] This realization has been shown to lead to high levels of stress, as documented in affected communities following the Deepwater Horizon oil spill.[38] Interestingly, some studies have shown that *despite* contamination—and perhaps because of a long-standing history and persistence as opposed to an acute episode—some places, such as the Anacostia River in the Washington, D.C., area continue to engender positive associations and attachment among those who live along its shores.[39]

Anthropologists, including Anthony Oliver-Smith, have observed that after severe disasters, which may physically destroy whole communities, people need to grieve in personally and culturally meaningful ways, not only for what may be loss of human and animal life but also for built structures and natural sites.[40] While what we may characterize as actual loss of place is clearly traumatic, potentially traumatizing changes in *sense* of self and place, or in personal attachment to place, can precipitate the same need to mourn and perhaps seek some form of resolution—the elusive experience of closure. For example, Gregory Button's report on long-term effects of the *Exxon Valdez* oil spill, a disaster that profoundly affected Alaska's Prince William Sound following the wrecked tanker's massive release of oil, found that many people continued to experience "depression, loss of sleep, and other symptoms of bereavement, though no human lives had been lost."[41]

In another example, Simona Perry documented feelings of heartfelt grief expressed by persons in her study who felt they were being adversely affected by shale gas development, regardless of their opinion concerning its overall economic benefit. She notes how her study participants spoke of feeling that their attachments to place—land, water, and family farms—were at risk of being lost due to changes precipitated by the experience of living near fracking sites. Perry quotes one resident who passionately states, "It feels like we're losing our love. The things we love the most may be taken away."[42] Much like what I observed in my ethnographic fieldwork among Balinese disaster victims, Oliver-Smith's work has shown that loss of place, both actual and symbolic, as well as in the relocation of people from places to which they have strong personal and collective attachment, may be deeply traumatic. As with the Balinese transmigrants in my Fulbright study in Indonesia, this may be so even when resettlement follows a disaster that makes the old places uninhabitable.[43]

Numerous studies of technological disaster—often involving at least *invisible* substances (e.g., chemicals or radiation) that pose real threats to health—document a shift not only in our overall sense of the world but also in the ability to physically sense a world changed by contamination. Despite the possible lack of immediate, sensory confirmation that something may be wrong, knowing (or even simply believing) that one may be affected by contamination through water, air, or food allows for fear and dread. For example, the effects of stress among residents near the Three Mile Island nuclear reactor persisted long after the disaster. A survey conducted several years later chronicled symptoms such as "'psychic numbing,' hopelessness, anxiety, feelings of being trapped by the situation and a lack of peace of mind."[44]

Considering our emergent understanding of the dimensions of place attachment, for example, in light of disaster studies, we may think of different places as existing on a continuum from the potentially therapeutic to the essentially pathogenic or disease causing. A therapeutic landscape, like a therapeutic community, with which people have a beneficial, supportive attachment necessarily represents only one dimension of our possible relationship with place. Geographer Yi-Fu Tuan referred to this kind of relationship with place as *topophilia*, the basis for positive affective attachment between person and place born of comfort and subjective well-being. A landscape of fear—or what Tuan described by way of his notion of *topophobia*, which refers to both psychological states and tangible environments—establishes an essentially negative or at least ambivalent relationship between people and place that may ultimately induce anxiety, dread, and depression.[45] In her work on the emotional development of children, Louise Chawla suggests that the places we inhabit have the potential for either light or darkness. There is always a "shadow side" to this relationship.[46]

While Chawla's concept evokes the relative darkness, it nevertheless opens the possibility of change in our relationship to place. The relationship is dynamic.[47] As Kroll-Smith and Couch explain, "contamination of a community's air, water or soil, for example, is likely to generate a high degree of uncertainty and confusion."[48] Essential questions are likely to remain unanswered: "What part of our environment is contaminated?" "Is it really dangerous?" "How should I behave . . . ?" Similarly, in speaking to the aforementioned "new species of trouble" of human-induced disaster, Kai Erikson suggests that toxic disasters "violate all the rules of plot"; although they may have clearly defined beginnings, invisible contaminants typically remain in peoples' bodies and in their environments.[49] In this way, this story, the narrative of contamination does not (indeed cannot) end.

When the rules of plot that otherwise narratively organize our senses of self are violated in this way, we may become stuck in a state of "ontological uncertainty." As described by Anthony Giddens, the ontologically secure individual is confident in the "continuity of their self-identity and in the constancy of the surrounding social and material environments of action." Such a person generally trusts that both persons and places that they encounter are consistent and reliable. As Giddens argues, however, "when routines are shattered—for whatever reason—anxieties come flooding in, and even very firmly founded aspects of the personality of the individual may become stripped away and altered."[50] Not surprisingly, this person becomes unmoored, lacking trust and confidence. Further, as was often invoked in the accounts of those interviewed for this project, being in a state of ontological uncertainty is exhausting. Many spoke of the physical and psychological energy it took to deal with and worry about their source and use of water. They waited—literally—and wondered if some sense of certainty or at least normality would ever again characterize their familiar routines.

In his study of community response to toxic exposure, Michael Edelstein suggests that such exposure affects different dimensions of lived experience in either or both "lifestyle" and "lifescape." On the one hand, *lifestyle* encompasses an individual's or group's "normal" set of behaviors that can be observed as patterned activities and the relationships between people, as well as people and the places that sustain them. On the other, *lifescape* is analytically intended to capture both individual and shared interpretive frameworks and assumptions that are used to understand the world and forecasting probabilities in it. Edelstein asserts that toxic exposure, among other disaster-initiated life experiences, may be fundamentally disruptive to these models of and for the world. For many people, such models may entail beliefs such as "humans hold dominion over nature, people control their own destiny, technology and science are progressive forces, environmental risks are acceptable, people get what they deserve, experts know best, the marketplace is self-regulating, one's home is one's castle, and government exists to help."[51]

Defining what some anthropologists now refer to as *precarity*, Andrea Muehlebach suggests that this analytical notion may serve as "a shorthand for those of us documenting the multiple forms of nightmarish dispossession and injury that our age entails."[52] As Tuan describes in his treatise on landscapes of fear, disaster experiences may provoke a "sickening sense of the dissolution of the known world" or what Katie Stewart refers to in her own examination of precarity as an "unworlding."[53]

Barbara Brown and Douglas Perkins discuss varied sources of disruptions in place attachment that include extreme situations such as those described in Kai Erikson's work on the Buffalo Creek flood. In this case the severity, wide-ranging impact, and apparent capriciousness of the disastrous flood "violated residents' assumptions about the world" where homes that were "formerly understood as bastions of safety, became weapons, trapping some neighbors in the flood, dismembering others."[54] So severe was the disaster that survivors had to try "to reconcile their religious belief that justice comes to the righteous with the fact that many flood victims were devoutly religious."[55]

Working in post-Chernobyl Russia and Ukraine some twenty years after the event, Patricia Abbott and her coauthors found that residents affected by this wide-ranging, traumatic event, which so utterly disrupted the lifescapes, saw their world and lives as having two time frames: before Chernobyl and after Chernobyl.[56] The nuclear disaster manifested genuine biographical disruption in the form of a far-reaching break in the lives of those affected with strong parallels to potential changes in sense of self as well as possible behavioral responses expressed in illness narratives at the onset of chronic disease. Those who survived the Chernobyl nuclear disaster lived as persons in a world made different for them. They could no longer see basic elements of their world or themselves as the same. Might this be true of those who experienced the Freedom Industries chemical release? At the least, many people cannot ignore the source, quality, and, ultimately, the vulnerability of their water. For some like Becky Park, the watershed moment when many affected by the water crisis came to think more critically about an indispensable but taken-for-granted resource like water was in fact a positive development for the very fact that the world was now seen differently, as though the scales had fallen from people's eyes.

Though seeing the world in a new light, taking on transformed mental perception of the way the world works, can be a valuable change, sense perception may not always be a reliable source for vital information. In direct reference to the Chernobyl disaster but applicable to contemporary experience in an industrial age that includes nuclear radiation as well as pervasive toxic chemicals, Ulrich Beck speaks to how we now experience a world essentially "unchanged for our senses, behind which a hidden contamination and danger occurred that was closed to our view—indeed, to our entire awareness. . . . The threat of the world behind the world remains completely inaccessible to our senses." Beck refers to this state of affairs as a "doubling of the world." As he goes on to assert, "Not just in the nuclear age, but with the industrial universalization of chemical poisons in the air, the water, and foodstuffs as well, our relation to reality has been fundamentally transformed."[57]

In her heart-rending application of Beck's approach, Sharon Stephens taught me that "'doubling' occurs in the broadening divide between the perceptual cues and forms of knowledge adequate for everyday life, and those forms required to grasp and respond" to such dangers as radiation. Specifically, in her work following the Chernobyl disaster among the Sami people, Sharon found that this doubling—together with necessary changes in both the Samis' production and consumption patterns as a result of contamination of their reindeer herds—required a painful and dislocating renegotiation of forms of individual and cultural identities once taken for granted, as well as their relationship with place. These changes are captured in statements such as this from one of her Sami informants: "It seems that things have become strange and make-believe. You see with your eyes the same mountains and lakes, the same herds, but you know that there is something dangerous, something invisible, that can harm your children, that you can't see or touch or smell. Your hands keep doing the work, but your head worries about the future."[58]

Perhaps recognizing the unsettling power of such doubling, media coverage of disaster has strategically made purposeful use of images that conjure disjuncture between real and apparent worlds. In his analysis of how toxic waste became an icon and mass issue through examining ways in which events such as the Love Canal tragedy—entailing contamination of a community by an abandoned waste dump previously operated by Hooker Chemical near Buffalo, New York—Andrew Szasz shows how media reports used visuals that seemed to signify "normalcy" but instead undermined or reversed it or signified the opposite. This could be accomplished through tactics like voice-over narration such as we hear in Szasz's description of one report where a boy is seen bicycling along a quiet suburban street while a narrator says, "There have been instances of birth defects and miscarriages among families," or when camera shots of "people's backyards, lawns, and swimming pools are followed by pictures of holes in the ground filled with ominously colorful soups of liquid chemicals."[59]

## Responses

I have spoken of ways in which disaster may affect people and provided some suggestion of reactions to this experience, but I have not yet attended at length to the question of how people willfully respond—including how in-cipient, collective action might take shape. A number of studies have shown that people, not only activists, who experience toxic contamination may shift personal conception of "the environment" as something discrete and distant "out there" to an "ambient" one of which they are a part. This was seen among Hungarian environmentalists, for example, after the Chernobyl disaster ex-panded the scope of their environmentalism beyond traditional concerns for flora and fauna by demonstrating how the home, workplace, and everyday spaces of community were also sites of vulnerability to environmental risk.[60]

Just as we have seen differential outcomes with respect to impacts of disaster on diverse communities, households, and persons being shaped by preexisting social conditions in what is then identifiable as a kind of *continuity* between situations before and after a disaster-precipitated crisis, these same conditions variously enable and limit possibilities for collective and individual responses at the time of and following a disaster.[61] In research in Louisiana's so-called Chemical Alley, Merrill Singer describes what he terms the "toxic frustration" of residents of an area heavily dependent on the chemical indus-try—much like the Chemical Valley of West Virginia. Within the larger context of this dependency and with this frustration tempering people's will to action, "on a day-to-day basis, they usually avoid thinking about . . . [environmen-tal health risks] too much unless they are prompted by specific events (like an oil spill, a chemical plant explosion, an unexplainable illness of a child, a sudden upturn in the foul taste of the local drinking water, the questions of a researcher, or similar events large and small)."[62]

Importantly, Singer explains that such debilitating frustration does not somehow exist in experiential isolation from other elements of one's social life but rather is "a component of a wider and deeper sense of enduring dis-appointment, social injury, and political economic marginalization."[63] In the post-Chernobyl landscape, Patricia Abbott and her coauthors similarly found deep feelings of individual resignation as well as a pronounced absence of po-litical mobilization.[64] That is to say, this state of frustration can be an *obstacle* to citizen action. This absence defies common sociological understanding re-garding what is normally a reliable impetus for social movement lying in the wake of the loss of legitimacy of government and important social institutions. Picou and his coauthors have found that, as "'systematic events' that permeate community social structure," disasters can also serve as *catalysts* for action even while they may be variously constraining in their own right.[65] Much of the literature on the contamination experience compellingly explores how what is often identified by those affected as a personal and collective—even willful—ignorance about both the presence and possible impact of toxic substances in their environment becomes disrupted and slowly replaced by attentiveness and emerging understanding. This shift is when individuals and even entire groups of people (often in specific neighborhoods) begin linking their places of work or residence with the prevalence of particular pathologies and connect what might be otherwise taken as an individual problem with those experienced by others.

People come to see patterns and to find these patterns meaningful. Investigating the experience of the community of Woburn, Massachusetts, as residents began to recognize a pattern of disease—an exceptional cluster of cases of leukemia—in their midst, Phil Brown and Edwin Mikkelsen identified and then named this process "popular epidemiology."[66] What might begin as happenstance discovery takes shape as an earnest, active course of self-education in such professional fields as public health. In the stories of those affected by the West Virginia water crisis, we hear how people thought carefully about possible differences in the apparent quality of the water, such as how it might be more harmful at higher temperatures. Some collected water samples in discarded jars as a way to keep track of possible changes and to document these trends for themselves in the present as much as for any future purpose. One woman spoke of her "obsessive" compulsion to collect such samples in little Tupperware containers over the first two weeks of the disaster. Another documented detail of the unfolding events including instructions from officials and her own household's timeline of decisions for following them in a day planner. Many spoke of a need to spontaneously share with others, often while waiting for water at distribution centers or in lines at stores, the details of their exposure experiences and thoughts about what to do going forward.

Patricia Abbott and her coauthors note that after an initial period of disruption following a disaster, or a realization of possible toxic exposure, many affected communities become politically mobilized.[67] These mobilizations are often founded on hard-won knowledge and skills learned while pursuing popular epidemiological explanations for their shared suffering. Through new tactics of information production and sharing at the grassroots level, Willow and Wylie describe how communities have organized virtual "bucket brigades."[68] This has been done in the context of citizen-based water quality monitoring, online mapping, and database sharing established by so-called non-experts in an attempt to fill gaps in public knowledge while documenting true hazards, risks, and environmental health impacts of, in their study, hydraulic fracturing.

Just as different dynamics may heighten the vulnerability of certain individuals and populations to the effects of disaster and constrain agency, so too may related factors provide post-disaster challenges to action—including organizing collectively. As suggested earlier, Singer's study in the Chemical Alley of Louisiana describes how the sheer economic and political power of the chemical industry intimidates local citizens and mitigates serious opposition.[69] While inaction in the face of such a constraining influence may be understood as wholly the product of outright fear—akin to that for a schoolyard bully—it is also strongly tied to conditions of poverty. Given the economic needs of individuals and communities in depressed areas with few other opportunities, chemical companies wield commanding threats like relocation in a place-based competition for what many communities take to be principal economic engines of job creation, that is, heavy industry.[70]

In the book *Ecology of Fear*, urban theorist and historian Mike Davis chronicles how Los Angeles, through the deceptive and self-serving marketing and sale of both real and imagined place by its boosters, has put itself in harm's way by building on fault lines, paving riverbeds, and turning floodplains into industrial zones, thereby transgressing environmental common sense. His account conveys tales of manufactured risk and human-constructed disaster. Davis describes how myths and misconceptions promulgated by powerful, vested interests purposefully and effectively shield people from having to acknowledge real dangers and their sources. In Southern California this includes decades of fire suppression in expensive suburbs accompanied by toxic rumors of arsonists from the city's inner city that allow wealthy suburbanites to (unproductively) focus their uncertainty and fear on a racial other. Disaster is repackaged as class struggle.[71]

Similarly, in West Virginia we have an abundance of cases, including the Freedom Industries chemical release into the Elk River, where common sense is subverted to special interest and human-created risks lay the framework for predictable disaster. Anthropologist Roy Rappaport would describe the crises that result from this subversion as being the sociocultural equivalent of an iatrogenic disorder in the realm of medicine; he compared them to situations where those responsible for providing care to others (i.e., a physician or *iatros* in Greek) instead witlessly through poor judgment or willingly with malice cause trauma instead.[72] Davis's use of the biological concept of an "ecology of fear" to name what he documents historically and observes on the ground in Los Angeles can inspire us to think about how collective fear can significantly affect behavior and how behavior thus shaped can substantially impact our shared environment and lives within it.

In biology the ecology of fear describes how predator species (e.g., wolves) provide a source of fear within prey species (e.g., elk) that face the threat of predation and thus are motivated to pursue particular kinds of foraging behavior that in the absence of predators would be quite different. The differences in foraging found in the absence of predators can lead to substantially different environmental conditions within the prey's habitat. At least in ecological science, fear experienced in the context of such predator-prey relationships can help maintain a sustainable order that scientists would characterize as mutually beneficial over time. Davis, however, is not using the ecology of fear as an intact concept outside the field of biology. Rather, he uses it as a trope that demonstrates how fear can be used by powerful interests to create not a mutually beneficial order but instead a disorder that purposefully shapes the social and physical landscape to suit self-interested purposes at the expense of public well-being. In such a context, citizens may be misdirected from real hazards in their environment.

Beyond overt challenges presented by purposefully elicited and misdirecting fear, Melissa Checker describes how everyday nuisances, such as the odd chemical smells periodically wafting over Charleston, West Virginia, can and do go unlinked by residents to local health concerns and problems. Checker explains that many residents in her studies told her that they were simply "too busy trying to live" in conditions of household economic scarcity to take the time and energy that they imagined would be necessary to trace any such connection and address their latent concerns.[73] Combined with distracted busyness, a loss of trust and goodwill in governmental agencies and potentially other organizational bodies meant to protect public welfare—as a result of their real or imagined recreancy—creates a potentially dangerous corrosiveness and failure to act.

As Simona Perry observes in the largely rural setting of her work, while the disruptive impact of a disaster or otherwise rapidly changed environment led people to begin to "articulate what the land, water, soil, rivers, wildlife, neighbors, families and sense of community truly means to them," it also engendered distrust in decision-making authorities.[74] Born of this distrust and now threatened by—or at least unsure about—a physical and social world that once served as a meaningful source of identity and point of attachment, loss of a sense of personal control can lead parents to a debilitating state of powerlessness in their perceived abilities to protect their children from harm. This agonizing disempowerment is captured in the tearful words of one of Willow's consultants who speaks as a mother raising her child in the shadow of fracking rigs: This is "the first time ever that I felt I had absolutely no control of keeping my child safe. Cause they took it away from me. The governor took it away from me, the ODNR [Ohio Department of Natural Resources] took it away from me, and [the energy company] took it away from me. And that is my job, to keep my child safe."[75]

In such a state, many whose worlds have "doubled," shifted into a kind of symbolic inversion wherein what remains familiar also becomes somehow strange, whose self-identities are threatened by the experience of ontological uncertainty, and whose homes no longer provide a sense of comfort and safety, find themselves thinking about moving away. As Janet Fitchen suggests based on her research on toxic contamination in light of cultural meanings of "home" in America, "a critically important function of the home is the sense of security that it is supposed to offer," something that is intensified—just as is its opposite sense of insecurity and fear following toxic exposure—when occupied by children.[76] When environmental toxins enter residential environments, Americans (and presumably others) perceive a multitude of threats beyond possible health risks and financial losses given that such contamination represents a violation of important cultural institutions of home and homeownership. This contamination is not restricted to home as physical structure but may effectively poison this potent symbol of familial integrity and the relationship that people have with that symbol.

Of course, many factors affect a family or household's ability to pull up stakes. Among them are financial considerations that include leaving existing or trying to find new jobs, lack of savings, and homes that may be difficult to sell in areas stigmatized through sociospatial contamination associated with a prominent technological disaster—these were all factors that figured prominently in the narratives of those interviewed in our project. In situations like this, as Rappaport posited, it is possible that "information concerning physical events, rather than the events themselves, act as stimuli" such that "physical events need not have yet occurred for there to be significant effects."[77] Uncertainty alone as a kind of information itself can be enough to dramatically shape decision-making behavior of both those affected and those unaffected by something like the chemical release in Charleston.

For many unable to relocate, however, the feeling of being stuck or that there is no way out from a place they have come to distrust or fear leads to prolonged stress and compromised mental and potentially physical health.[78] That said, Brown and Mikkelsen have found, as they observed among affected residents in Woburn, including leukemia victims and their families, most people choose to remain despite their exposure to known hazards.[79] The same has been true for people here in West Virginia—they stay because they have come to the realization that there may be no truly safe places to which they can retreat. As we heard from Anita Edmonds in chapter 6, "Probably this could happen about anywhere. . . . I don't feel like it's any more dangerous to live here than it would anywhere else. I kinda feel like it would be safer to live here." Might this be a matter of "the devil you know"?

## On Making and Remaking Community

Jim Hatfield opens part III, the theme of which is the making and remaking of community in the release's aftermath, from the perspective of a chemical engineer with a long tenure in the Chemical Valley. Hatfield describes profound failures of state agencies and the private water utility to prepare for the eventuality of a chemical release of the magnitude seen with Freedom Industries. His chapter picks up a narrative thread about local activism introduced by his wife, Trish, at the end of part I. Starting with the experience of a local grassroots nonprofit responding to the needs of persons of the most limited means who in some cases had little choice but to continue to use contaminated water, Hatfield explores activism manifest in a multitude of ways through existing commitments to community by established organizations as well as emergent individual and collective efforts, shaped by transformative experiences ignited by acute water contamination, to make changes in status quo relationships between government, industry, and citizens of the state.

Gabe Schwartzman's chapter presents how he came to the Appalachian region from California, before the January 2014 disaster, with the intent to study a long history of water crises throughout the area. The contamination provided an unexpected focal point for his work, which, among other things, documents how suffering in the context of disaster is largely attributable to systemic inequality with those most vulnerable disproportionately exposed. The findings of his collaborative research are presented on a purposefully designed website that combines varied accounts of living with contaminated water together with maps and interactive graphics in a digital storytelling platform that offers another point of entrée into varied experiential dimensions of the water crisis. The site offers a systematic overview of privatization in public water utilities within affected communities that has served policy advocates for safe water systems.

Similarly, Laura Allen—a native documentarian and independent radio producer who has worked with West Virginia Public Radio—describes her eventual role in a graduate seminar conducted with Eric Lassiter, in which students worked to make an audio documentary of the water crisis. This work began with her role helping organize responses by local charities while simultaneously "gathering tape," as she put it, in the summer of 2014 to capture accounts of people who are all too often overlooked in mainstream media coverage and to document the economic impact on marginalized peoples. While Allen finds the script of the water crisis essentially familiar in terms of governmental and industry response, the particulars of this place and the relationship of people to it are a critical distinction where the state of West Virginia becomes, to Allen, a character in itself.

In closing part III, Hatfield returns to set up our epilogue, which, broadly considering "outcomes," focuses on the terms of the collaborative nature of our project. Specifically, Hatfield provides us a sense of where we are now with regards to the chemical release and of responses to it at different levels—from the individual to collective—by varied actors. We are given an overview of both achievements and shortfalls in efforts by local activists to find some justice and ultimately safeguard the water supply.

## Notes

1. Pamela Abbott, Claire Wallace, and Matthias Beck, "Chernobyl: Living with Risk and Uncertainty," *Health, Risk and Society* 8, no. 2 (2006): 105–21.
2. Ben McMahan, "Disaster: Provocation," Cultural Anthropology Online, accessed March 3, 2015, http://www.culanth.org/fieldsights/144-disaster-provocation.
3. Doug Henry, "Anthropological Contributions to the Study of Disasters," in *Disciplines, Disasters and Emergency Management: The Convergence of Concepts Issues and Trends from the Research Literature*, ed. D. McEntire and W. Blanchard (Emittsburg, MD: Federal Emergency Management Agency, 2005), 8, emphasis added.
4. Brian A. Hoey, "Creating Healthy Community in the Postindustrial City," in *Recovery, Renewal, Reclaiming: Anthropological Research toward Healing*, ed. Lindsey King (Knoxville, TN: NewFound Press, 2015).
5. J. S. Kroll-Smith and Stephen R. Couch, "Technological Hazards," in *International Handbook of Traumatic Stress Syndromes*, ed. John P. Wilson and Beverley Raphael (New York: Springer, 1993).
6. Anthony Oliver-Smith, "Anthropological Research on Hazards and Disasters," *Annual Review of Anthropology*, 1996; Anthony Oliver-Smith and Susanna M. Hoffman, "Introduction: Why Anthropologists Should Study Disasters," in *Catastrophe and Culture: The Anthropology of Disaster*, ed. Susannah M. Hoffman and Anthony Oliver-Smith (Santa Fe: School of American Research Press, 2002).
7. William I. Torry, "Morality and Harm: Hindu Peasant Adjustments to Famines," *Social Science Information* 25, no. 1 (1986): 126.
8. Cf. Francis O. Adeola and J. S. Picou, "Social Capital and the Mental Health Impacts of Hurricane Katrina: Assessing Long-Term Patterns of Psychosocial Distress," *International Journal of Mass Emergencies and Disasters* 32, no. 1 (2014): 121–56.
9. Piers Blaikie, Terry Cannon, Ian Davis, and Benjamin Wisner, *At Risk: Natural Hazards, People's Vulnerability, and Disasters*, 2nd ed. (London, New York: Routledge, 2004); Susan L. Cutter, Bryan J. Boruff, and W. L. Shirley, "Social

Vulnerability to Environmental Hazards," *Social Science Quarterly* 84, no. 2 (2003): 242–61.

10. W. Neil Adger, Terry Hughes, Carl Folke, Stephen Carpenter, and Johan Rockström, "Social-Ecological Resilience to Coastal Disasters," *Science* 309, no. 5737 (2005): 1036.

11. P. van Arsdale, "The Ecology of Survival in Sudan's Periphery: Short-Term Tactics and Long-term Strategies," *Africa Today* 36, no. 3/4 (1989): 65–78.

12. Adeola and Picou, "Social Capital," 2.

13. Duane A. Gill, Liesel A. Ritchie, Steven J. Picou, Jennifer Langhinrichsen-Rohling, Michael A. Long, and Jessica W. Shenesey, "The Exxon and BP Oil Spills: A Comparison of Psychosocial Impacts," *Natural Hazards* 74, no. 3 (2014): 1911–32.

14. Marc Fried, "Continuities and Discontinuities of Place," *Journal of Environmental Psychology* 20, no. 3 (2000): 193–205.

15. E.g., see Gregory V. Button, "'What You Don't Know Can't Hurt You': The Right to Know and the Shetland Island Oil Spill," *Human Ecology* 23, no. 2 (1995): 241–57, regarding access to and control of information.

16. Krista M. Harper, "Chernobyl Stories and Anthropological Shock in Hungary," *Anthropological Quarterly* 74, no. 3 (2001): 114–23; cf. Melissa Checker, "'But I Know It's True': Environmental Risk Assessment, Justice, and Anthropology," *Human Organization* 66, no. 2 (2007): 119.

17. Anna J. Willow, "The New Politics of Environmental Degradation: Un/Expected Landscapes of Disempowerment and Vulnerability," *Journal of Political Ecology* 21 (2014): 248.

18. Kroll-Smith and Couch, "Technological Hazards"; Michael R. Edelstein, *Contaminated Communities: Coping with Residential Toxic Exposure*, 2nd ed. (Boulder, CO: Westview Press, 2004).

19. Kai Erikson, *Everything in Its Path: Destruction of Community in the Buffalo Creek Flood* (New York: Simon and Schuster, 1976).

20. Gill et al., "Exxon," 14.

21. William R. Freudenburg, "Contamination, Corrosion and the Social Order: An Overview," *Current Sociology* 45, no. 3 (1997): 19–39; cf. J. S. Picou, Brent K. Marshall, and Duane A. Gill, "Disaster, Litigation, and the Corrosive Community," *Social Forces* 82, no. 4 (2004): 1493–1522.

22. Picou et al., "Disaster," 1494; cf. Edelstein, *Contaminated Communities*.

23. Picou et al., "Disaster," 1496.

24. Peter C. Little, *Toxic Town: IBM, Pollution, and Industrial Risks* (New York: New York University Press, 2014); Peter C. Little, "Another Angle on Pollution Experience: Toward an Anthropology of the Emotional Ecology of Risk Mitigation," *Ethos* 40, no. 4 (2012): 431–52.

25. Little finds that for most people, the official contamination response of installing easily perceptible fans now serves as both visual and auditory reminders of possible exposure to toxins—or what Little appropriately refers to as, "mnemonic devices" for residents to regularly recall their own real or imagined contamination. Little, *Toxic Town*.

26. Martha A. Ballman, interview with Jim Hatfield, August 26, 2014.

27. Arthur Kleinman and Joan Kleinman, "The Appeal of Experience, The Dismay of Images: Cultural Appropriations of Suffering in Our Times," *Daedalus* 125, no. 1 (1996): 1–23.

28. Gregg Mitman, Michelle Murphy, and Christopher C. Sellers, *Landscapes of Exposure: Knowledge and Illness in Modern Environments* (Chicago: University of Chicago Press, 2004), 11.

29. Cf. Javier Auyero and Débora Swistun, *Flammable: Environmental Suffering in an Argentine Shantytown* (New York: Oxford University Press, 2009); Merrill Singer, "Down Cancer Alley: The Lived Experience of Health and Environmental Suffering in Louisiana's Chemical Corridor," *Medical Anthropology Quarterly* 25, no. 2 (2011): 141–63.

30. Cf. Crystal Adams et al., "Disentangling the Exposure Experience: The Roles of Community Context and Report-Back of Environmental Exposure Data," *Journal of Health and Social Behavior* 52, no. 2 (2011): 180–96; Rebecca G. Altman, Rachel Morello-Frosch, Julia Green Brody, Ruthann Rudel, Phil Brown, and Mara Averick, "Pollution Comes Home and Gets Personal: Women's Experience of Household Chemical Exposure," *Journal of Health and Social Behavior* 49, no. 4 (2008): 417–35.

31. Peter J. Guarnaccia, Glorisa Canino, Maritza Robio-Stipec, and Milagros Bravo, "The Prevalence of Ataques de Nervios in the Puerto Rico Disaster Study: The Role of Culture in Psychiatric Epidemiology," *Journal of Nervous and Mental Disease* 181, no. 3 (1993): 157–65.

32. Joshua Reno, "Beyond Risk: Emplacement and the Production of Environmental Evidence," *American Ethnologist* 38, no. 3 (2011): 521. See also Linda Green, "Lived Lives and Social Suffering: Problems and Concerns in Medical Anthropology," *Medical Anthropology Quarterly* 12, no. 1 (1998): 3–7.

33. Anna J. Willow and Sara Wylie, "Politics, Ecology, and the New Anthropology of Energy: Exploring the Emerging Frontiers of Hydraulic Fracking," *Journal of Political Ecology* 21 (2014): 226; cf. Brian A. Hoey, "Place for Personhood: Individual and Local Character in Lifestyle Migration," *City and Society* 22, no. 2 (2010): 237–61.

34. Little, "Another Angle," 446–47.

35. Hoey, "Creating Healthy Community."

36. J. S. Kroll-Smith and Stephen R. Couch, "What Is a Disaster? An Ecological-Symbolic Approach to Resolving the Definitional Debate," *International Journal of Mass Emergencies and Disasters* 9, no. 3 (1991): 355–66; Kroll-Smith and Couch, "Technological Hazards"; Hoey, "Creating Healthy Community."

37. Auyero and Swistun, *Flammable*; Thomas D. Beamish, "Accumulating Trouble: Complex Organization, a Culture of Silence, and a Secret Spill," *Social Problems* 47, no. 4 (2000): 473–98; see also, for consideration of such watersheds in the lifecourse in the context of disruption in work life, Brian A. Hoey, *Opting for Elsewhere: Lifestyle Migration in the American Middle Class* (Nashville, TN: Vanderbilt University Press, 2016).

38. M. R. Lee and T. C. Blanchard, "Community Attachment and Negative Affective States in the Context of the BP Deepwater Horizon Disaster," *American Behavioral Scientist* 56, no. 1 (2012) 24–47.

39. Brett Williams, "A River Runs through Us," *American Anthropologist* 103, no. 2 (2001): 409–31.

40. Oliver-Smith, "Anthropological Research," 309. Cf. Stewart Kirsch, "Lost Worlds: Environmental Disaster, 'Culture Loss,' and the Law," *Current Anthropology* 42, no. 2 (2001): 167–98.

41. Gregory V. Button, *Disaster Culture: Knowledge and Uncertainty in the Wake of Human and Environmental Catastrophe* (Walnut Creek, CA: Left Coast Press, 2010), 47.

42. Simona L. Perry, "Development, Land Use, and Collective Trauma: The Marcellus Shale Gas Boom in Rural Pennsylvania," *Culture, Agriculture, Food and Environment* 34, no. 1 (2012): 88.

43. Hoey, "Nationalism in Indonesia"; Anthony Oliver-Smith, "Here There Is Life: The Social and Cultural Dynamics of Successful Resistance to Resettlement in Post-disaster Peru," in *Involuntary Migration and Resettlement: The Problems and Responses of Dislocated People*, ed. Art Hansen and Anthony Oliver-Smith (Boulder, CO: Westview Press, 1982).

44. Roger E. Kasperson and K. D. Pijawka, "Societal Response to Hazards and Major Hazard Events: Comparing Natural and Technological Hazards," *Public Administration Review* 45 (1985): 14.

45. Yi-Fu Tuan, *Topophilia: A Study of Environmental Perception, Attitudes, and Values* (Englewood Cliffs, NJ: Prentice-Hall, 1974).

46. Louise Chawla, "Childhood Place Attachments," in *Place Attachment*, ed. Irwin Altman and Setha Low (Boston, MA: Springer, 1992).

47. Hoey, "Creating Healthy Community."

48. Kroll-Smith and Couch, "What Is a Disaster?" 363.

49. Kai Erikson, "A New Species of Trouble," in *Communities at Risk: Collective Responses to Technological Hazard*, ed. Stephen R. Couch and J. S. Kroll-Smith (New York: P. Lang, 1991), 148.

50. Anthony Giddens, *The Consequences of Modernity* (Cambridge: Polity, 1990), 92, 98.

51. Edelstein, *Contaminated Communities*, 24.

52. Andrea Muehlebach, "On Precariousness and the Ethical Imagination: The Year 2012 in Sociocultural Anthropology," *American Anthropologist* 115, no. 2 (2013): 298.

53. Tuan, *Landscapes*, 69; Kathleen Stewart, "Precarity's Forms," *Cultural Anthropology* 27, no. 3 (2012): 518–25.

54. Barbara B. Brown and Douglas D. Perkins, "Disruptions in Place Attachment," in *Place Attachment*, ed. Irwin Altman and Setha Low (Boston, MA: Springer, 1992), 291. Cf. Edelstein, *Contaminated Communities*, on the notion of "inversion of the home."

55. Brown and Perkins "Disruptions in Place," 291.

56. Abbott et al., "Chernobyl," 111.

57. Ulrich Beck, "The Anthropological Shock: Chernobyl and the Contours of the Risk Society," *Berkeley Journal of Sociology* 32 (1987): 153–65; cf. Harper, "Chernobyl Stories," 154, 155.

58. Sharon Stephens, "The 'Cultural Fallout' of Chernobyl Radiation in Norwegian Sami Regions: Implications for Children," in *Children and the Politics of Culture*, ed. Sharon Stephens (Princeton, NJ: Princeton University Press, 1995), 294, 298.

59. Andrew Szasz, *Ecopopulism: Toxic Waste and the Movement for Environmental Justice, Social Movements, Protest, and Contention* (Minneapolis: University of Minnesota Press, 1994), 43.

60. Harper, "Chernobyl Stories."

61. G. A. Kreps, "Disasters and the Social Order," *Sociological Theory* 3 (1985); Oliver-Smith, "Anthropological Research."

62. Singer, "Down Cancer Alley," 158.
63. Singer, 158.
64. Abbott et al., "Chernobyl."
65. Picou et al., "Disaster," 1495.
66. Phil Brown and Edwin J. Mikkelsen, *No Safe Place: Toxic Waste, Leukemia, and Community Action* (Berkeley: University of California Press, 1990).
67. Abbott et al., "Chernobyl"; cf. Oliver-Smith, "Anthropological Research."
68. Willow and Wylie, "Politics."
69. Singer, "Down Cancer Alley."
70. Brian A. Hoey, "Capitalizing on Distinctiveness: Creating West Virginia for a 'New Economy,'" *Journal of Appalachian Studies* 21, no. 1 (2015): 64–85.
71. Mike Davis, *Ecology of Fear: Los Angeles and the Imagination of Disaster* (New York: Metropolitan Books, 1998). See also how Auyero and Swistun speak to the ways that socially produced "risk frames" in the context of manipulative power may affect what people are able to see and not see, know and not know, Auyero and Swistun, *Flammable*.
72. Roy A. Rappaport, "Disorders of Our Own: A Conclusion," in *Diagnosing America: Anthropology and Public Engagement*, ed. Shepard Forman (Ann Arbor: University of Michigan Press, 1995), 235–94.
73. Checker, "'But I Know It's True,'" 116.
74. Perry, "Development," 90.
75. Willow, "New Politics," 248.
76. Janet M. Fitchen, "When Toxic Chemicals Pollute Residential Environments: The Cultural Meanings of Home and Homeownership," *Human Organization* 48, no. 4 (1989): 83.
77. Roy A. Rappaport, "Human Environment and the Notion of Impact," in *Life and Death Matters: Human Rights and the Environment at the End of the Millennium*, ed. B. R. Johnston (Walnut Creek, CA: AltaMira, 1994), 161; cf. Beck, "Anthropological Shock," 153.
78. Cf. Abbott et al., "Chernobyl"; Checker, "'But I Know It's True,'"; Little, *Toxic Town*.
79. Brown and Mikkelsen, *No Safe Place*.

Part III

**On
Making
and
Remaking
Community**

# CHAPTER 8

# Activism and Community

Jim Hatfield

David Chairez, deputy director of Step by Step, Inc., a regional grassroots nonprofit serving West Virginia, had a bird's-eye view of the extra trauma the crisis caused for those of meager means. Many people, he remembered, "just kept using the tap water because they didn't have any other choice. They had to prepare their food. They had to continue washing their clothes. They had to continue bathing themselves and . . . drinking the water. Some people drank the contaminated water because they didn't have the resources to do otherwise. Other people drank the water because they didn't have timely information about the dangers of doing so. And still other people drank it because they'd lived through hard times before. They've survived x, y, and z, and they believed they would survive this crisis, too. . . . Whatever the reason, the people we saw who suffered the most were families who had young children and infants. Imagine having to fix baby formula with contaminated water. And using that water to cook with, to feed your children, because restaurants were closed and grocery stores were picked clean. So many other businesses were closed, too, and many parents couldn't go to work. So now they couldn't make money, plus they're spending more money than usual. So when the next week comes along they don't have the money they need in order to buy potable water, to buy food, or to buy more formula. Now those families are in the hole twice as much as they would normally be when they were already just trying to make ends meet."[1]

For David, delivering potable water to people his organization served was a number one concern. But because the crisis affected people across the board, every economic class, every part of the community, such concerns were not limited to those like David and his organization. Shelda Martin, the physician introduced in Eric's chapter 1, for example, noted that "we have a lot of students and residents here [at Charleston Area Medical Center Memorial Hospital] who live in residency housing. And so my question—because I sit on the housing board for these guys—was 'Do they have water?'"[2] Marlene Price, who works for the West Virginia Department of Education, had similar concerns: "I had bottled water at home because I keep it for travel. . . . I had a couple of cases, but what about the people who didn't? And I became concerned about the elderly, the people who couldn't get out to get some water. And poor people who can't afford it. Immediately my family was calling and saying, 'Who's checking on so-and-so? Do we need to go by there? Let's run up and see if we can get water from Walmart or Sam's or something, [so] that we can buy several cases and just give it to people so they'll have it.'"[3]

People like David, Shelda, and Marlene, along with many organizations, emergency responders, and service providers, took on an important role in responding to the crisis and helping those who needed it most. But so did individual everyday people. Indeed, the desire to help came not only from state and federal agencies, not only from those employed in nonprofits and governmental agencies. As pivotal and essential as all of this help was, those affected also sought ways to help family, friends, and neighbors—one another—as the water crisis unfolded. "We see friends from other areas offering their homes and their showers," said Lucy Nelson, a local yoga instructor, "and just a lot of support."[4] Belinda Vance noted how neighbors went out of their way to check on each other: "The older lady that lives down from us," she remembered, "she's 82, and she's deaf and lives alone. [My husband and I] made sure that she had water and that she knew what was going on."[5]

Countless folks, it turns out, tracked down and delivered potable water. They provided transportation to showers and washing machines; they encouraged, consulted, commiserated, and just checked in. And they worried about those they couldn't help and those barely scraping by. All these acts are essential in times like these: though largely individual, these actions of a community in crisis, repeated over and over again, pull together strands of obligation and resolve. They create something larger than individuals, a tide of collective caring, something like an ocean tide created by the small but persuasive gravitational attraction of the moon on individual molecules that comprise the sea. These actions of caring were also new demands, demands added to almost every interviewee's daily rounds when the water company CEO cautioned that the water was undrinkable and the governor declared a public water emergency on the winter evening of January 9, 2014.

## An Engineer Turned Activist: On Who I Am and Where I'm Coming From

This chapter is about the commitment of people involved in the crisis to help others on scales both large and small. It focuses, in particular, on the development of collective action around the need—and guarantee—for clean water in our communities.

Soon after the chemical spill, I became involved with a newly formed group called Advocates for a Safe Water System (ASWS). My background is technical: I have undergraduate, masters, and PhD degrees in chemical engineering and a twenty-five-year career as a research scientist here in Chemical Valley. I worked—with great enjoyment, I might add—at Union Carbide's largest research center. The tractor beam that pulled me into an activist role was several newspaper articles in our local newspaper, the *Charleston Gazette*, mostly those written by reporter Ken Ward Jr. These articles revealed a primitive state of chemical instrumentation at the Elk River treatment plant and reported what, to my ears, had the ring of obfuscatory technobabble from some water company and state agency officials. How could this water plant, a chemical production unit in its own right, have operated in such a technology-deficient bubble, surrounded as it is in the Kanawha Valley by world-class chemical research and manufacturing facilities? From my vantage point, this technical deficiency on the part of the water company and the state was a central flaw that helped magnify the January 9 chemical spill into a nine-county, regional public drinking water disaster of national proportion. The event hurt many individuals and families economically, physically, and emotionally. Knowing this and then reading too many half-baked technical "explanations" delivered in cavalier and arrogant tones propelled me into the land of activism. In a short time, my own activist credo crystallized into this simple question: "What is the shared responsibility for the water crisis, and how can a similar crisis be avoided in the future?" This chapter explores how others found and enacted their own activisms.

## Activism, Part I

That most adults cannot extend themselves beyond caring for family, friends, and neighbors in the onslaught and aftermath of crises is hardly news. The demands of family, work, and other pressing obligations are daunting, and more than fully occupy most members of our society. Any crisis-inspired response, short- or long-lived, is a new activity with its own additional time requirement. Nevertheless, many interviewees were motivated to carve out their own "crisis response," something more than surviving and helping family and friends survive the crisis, important as that is, something that contributed to a broader understanding of what happened and, perhaps, the hope for some kind of change.

Charleston resident Becky Park felt motivated to do something more. Soon after the spill, she volunteered to help collect crisis-related statistics for the West Virginia Department of Health and Human Resources: "I think that twice I went down and worked for as many hours as I could," she said. "It might've been three or four hours at a time to call around West Virginia and run through, I think it was, like, fifteen or twenty pages of questions to ask people."[6]

More than a few residents helped with work like this and other research projects (doing oral history, for instance, and agreeing to be interviewed, as in our own project, was certainly a part of this). Other residents responded by writing about their experiences on internet blogs. Others wrote letters to the editor and op-eds for the local newspaper. And still others made YouTube videos, commentaries, and other documentaries. Austin Susman, a high school senior during the crisis, was out of school for almost three weeks courtesy of the contaminated water. It was time enough to join with three theater friends and create a music video about some consequences of the event. "There was a campaign started by members of our community called Turn Up the Tips," he remembered, "which basically meant that, you know, restaurant workers had been out of work for pretty close to a week when the [water use] ban was lifted, so they lost out on their pay. So it was just encouraging people in the Valley to tip a little bit more when they went out to eat in the weeks following the chemical spill."[7]

Austin continued, "myself, Daniel Calwell, and Caitlyn Moore—we do a lot of video work together—and so we started to talk about ideas of doing a video about the water crisis. We had a couple different concepts and we thought about mentioning the Turn Up the Tips thing, and it ended up being what we did our whole music video about. . . . There was already enough stuff going out on the Internet about being angry about the fact that [the crisis] happened. And we thought, let's do something a little more positive. . . . We got a *crazy* good response. And we didn't expect it. We had, I want to say in the first day, we had over a thousand hits on YouTube. And then we emailed the link out to a lot of local radio stations and news stations. I know Hoppy Kercheval [a local radio host] tweeted about us. . . . And then WOWK actually ran a piece about our video and we got interviewed. We got our story in the *Charleston Gazette* as well."

Acts like these helped create a rising tide of consciousness in the wake of the water crisis.

As it unfolded, many found themselves willing and able to participate more fully in crisis response. In this phase of collective public activity, a growing activism began to emerge. Some activisms dealt with the immediate fallout from the emergency, and others began to probe and address root causes and more systemic components of the crisis. A more full-fledged commitment to activism began to simmer. For some the passion to participate in community development and well-being was born on January 9; for others, a commitment that had been around for years found a new venue for expression. Take, for example, musician and teacher Paul Epstein.

Paul grew up in Bethlehem, Pennsylvania, and moved to the Mountain State in 1974. He had discovered a musical bent early, at least by second grade, when he started piano lessons. In his late teens, a guitar seemed more reasonable, and a fiddle was added when he moved to West Virginia. He spent eighteen years in West Virginia's rural Roane County. For part of that time he was a troubadour traveling Appalachian country roads with one or another of his bands playing traditional string music and bluegrass. Life was good, but with the arrival of his first child, the need for a steady income led Paul to his twenty-five-year career as an elementary school teacher. Until his retirement from teaching a few years ago, playing in musical bands took a backseat in Paul's life, but that musical life never completely disappeared.

In some ways, Paul feels that his country living experience helped deflect the blow of the water crisis. "Having lived in the country," he said, "having had wells and . . . [living] at different times in my life without running water, I wasn't really thrown too far off balance by the fact that I might have to live without running water for a week or so if that's what it came to."[8] For Paul the crisis began as a "hybrid experience" of sorts: inconvenience plus adventure. "I had to be . . . a hunter every day and look for water," he said. Naturally attuned to the ebb and flow of information and opinion in the community, he was concerned for those who didn't appreciate potential hazards of the contaminated water. He became impatient with some officials he felt were too submissive. "The mayor said, 'It's out of my hands,'" Paul remembered. "'There's nothing we can do at the local level. We have to wait for crews to come from outside.'" And others seemed too cautious. The governor "needed to make a decision, but he didn't. He tried to walk a fine line . . . and that got him in trouble."

The crisis moved from simmer to a boil for Paul as it became clearer that more open and vocal advocacy was necessary. And it awakened in him a response having to do with the larger environmental and human damage caused by things like mountaintop removal coal mining and fracking in the pursuit of natural gas reserves. In sum, it prompted "a strong personal response to the whole situation. . . . I wasn't upset by the immediate loss of water: I saw it as an accident. However, I did place blame for the accident; I did see that it was a preventable accident on the [part of the] Department of Environmental Protection [DEP], and our legislature, and the governor, and previous governors and legislatures . . . especially going back to Joe Manchin, who I felt weakened the DEP. . . . This spill . . . awakened within me a somewhat latent impulse to do environmental work and to advocate against mountaintop removal coal mining, the injustice homeowners are experiencing in the gas fracking fields . . . basically wanting government to do its job of regulating these industries and making sure that we're safe. . . . [I understand and appreciate] that there is a balance between industry and public safety, but the balance should err on the side of public safety."

So in the weeks and months after the spill, Paul "went to public meetings about the water crisis. I wanted to be informed and I also wanted to express my voice as a citizen, my anger at the DEP and the Governor for not having done a better job of regulating this industry—whether you call it the coal industry or the chemical industry. . . . I went to Earth Day in April to lobby some of the legislators on the water bill." And here Paul returned to his musical roots. "I volunteered as a musician," Paul said. And "I wrote a song. . . . Somewhere in there I had already begun having ideas and talking to people about what I could do to help the environmental movement in West Virginia. I realized that, as a musician, with a lot of ties to musicians and artists who are often being asked to provide entertainment or art to support environmental issues to help raise money. . . . I thought, 'Well maybe *there's* a way I could use my expertise to help.'"

Paul's efforts grew and other artists came together under an umbrella "project we ended up calling AWARE [Artists Working in Alliance to Restore the Environment]. So I focused on gathering up a group of advisors and organizing a fundraiser, which I eventually had in July. I raised about $5,000 and was able to distribute about $2,500 of it to the West Virginia Environmental Council and its member groups. So I feel good about that."

## Activism, Part II

Like Paul Epstein, others were moved by the water crisis to do something different. On the afternoon I interviewed Martha Ballman at her home, her two corgis seemed more focused on neighborhood comings and goings than our conversation about the water crisis. As we sat talking in the front yard garden, with the soothing sounds of wind chimes in the air, it became clear that Martha had thought at length about the crisis, its impact on the community and herself, and how it had pushed her farther down the path of citizen involvement. "This whole incident has created an awareness in, I think, all of us that it's up to us to do something. It's the first time in my life I've ever been actively an 'activist'—you know, show up at the capitol for the hearings [on clean water legislation] and go out with petitions and do things now. Our environment is sacred. We *have* to have clean water."[9]

Like Paul, Martha had also been thinking about the water company's role in the crisis and in the community: "I had no idea that West Virginia American Water . . . had bought out [so many] small community and municipal water systems. . . . So now they're too big to fail. . . . There's no accountability on the local level. It's not like the city council: You know who they are. You vote for them. If you have a gripe you can show up at city council. And you're a person, you have a voice. . . . There's got to be a better way to be heard, a more strategic way of letting them understand that there's people here they're affecting and we're aware and we're not happy about it. . . . I actually talked to the city planner in Charleston shortly after things happened and asked, 'What's the possibility now of Charleston and South Charleston and some of these municipalities taking back their water and running it?' And I think the citizens would be willing to pay even more for their water if they knew they'd get this infrastructure fixed. I mean we're all sick to death of water leaks."

Martha's comments revealed some of the human impact of the water company's "unaccounted-for" water running as high as 30 and 35 percent: for every gallon of treated water they produce, one-third is lost to system leaks as well as miscellaneous activities such as flushing water lines, back-flushing their massive filter beds regularly, and other process uses. (The industry standard is about 15 percent.) She continued in this vein: "I have no faith that those leaks are not affecting our water quality. . . . We're ignorant about the whole thing. We're ignorant about the pipes: Where are the pipes? What kind of pipes? How old are they? What's the health of these pipes?"

Martha abruptly turned her attention to improving the water condition. "I think the only way we're going to get better at this," she said, "is to be more educated about it and to be more diplomatic about it. And we need to be willing to devote a lot of energy and a lot of time. Not just for the moment. It's got to be a long haul thing or it's not worth doing. I called the city and I said, 'I will work for this. I will grant write. I will research. I would do whatever we need to do. I'll politic. But it's got to be for something I feel is worthwhile, I mean, it's really going to fix it and not just make noise.'"

Martha is the kind of person who walks the walk. Six months after her interview, a year after the chemical leak, she began spending more time advocating for safe water by diving into the crisis-born group mentioned earlier in this chapter, Advocates for a Safe Water System. At the time of this writing, her commitment has not waned; it's actually grown. The group's focus has expanded to a larger campaign, beyond simply improving the existing water system so it can face the next chemical spill without contaminating the public water supply—no small accomplishment in itself—to municipalizing the water utility. This larger goal coincides with Martha's own initial reaction to the crisis. Her time commitment has grown as she walks neighborhoods carrying petitions, shares her own crisis experience at public meetings, helps promote the group's agenda at public functions and has become a fixture in the ASWS organization.

---

Groups like ASWS were not the only organizations that worked for change. Many churches, like Charleston's Unitarian Universalist Congregation (UUC), took an active role in organizing relief, too, as well as serving as forums for critical dialogue about what could be done. With this in mind, one afternoon I interviewed Rose Edington, who, along with her husband, Mel Hoover, were about to finish their many years of service as co-ministers of UUC when the chemical spill happened. (Both are retired now.)

Rose, who grew up in nearby Saint Albans, is a long-term activist. Having lived much of her life in the Kanawha Valley, the so-called Chemical Valley, she has seen her share of chemical catastrophes. Her outlook is a balance of hope (based on her considerable experience with caring individuals, churches, and organizations) and skepticism (based on her observation of bureaucratic inertia and corporate influence and recklessness). But water issues are of particular concern to her because it's so significant to UUC members generally. On the day I interviewed her, I couldn't help but notice, for example, the recent September UUC newsletter on the coffee table. In bold letters it shouted: "Water is life." So I asked Rose to say a little more about UUC's connection to water before we focused on the water crisis.

"In the early 1980s," Rose explains, "the Unitarian Universalist Women's Federation called for a nationwide meeting. There was a whole lot of change going on in the faith because the Unitarian Universalist Association and the Universalist Church of America merged in 1961. . . . The women [gathered there] said, 'We need to figure out our role in this faith system.' . . . As part of the gathering they had a uniting water ceremony/ritual, and when the women went back to their churches they said we should do some kind of water communion every year in our congregation. So it spread from the women going to that national conference. . . . The focus, at least one Sunday in September, is on water. But because of our situation here [i.e., the water crisis], the whole month is devoted to water."[10] Rose continued to describe how, every spring, congregants are reminded to collect water samples during their summer outings and travels: "At the 'in-gathering' service [following Labor Day] we have everyone, all the family, come up and pour their water into a common bowl and share the significance of the water, where they went, what the water means to them."

For Rose, Mel, and their congregation, the water crisis inspired more than just discomfort and critique: it prompted the church to act as listening ears, helping hands, and distributors of water and money. Rose stated matter-of-factly: "Our church was very helpful . . . We talked with our district executive [of the association of congregations] and wrote up a little piece about what was going on here, and it got sent out through the church network. A lot of our congregations do this once-a-month plate collection that supports some kind of social justice. So a lot of plate collections were taken for us. And twenty-three thousand dollars came to UUC to deal with the water issues here from other UU's—mostly from the Appalachian region, but there were also contributions from Texas and California as more people heard about it . . . over two or three months. It was up to the congregation and the [new] interim minister [who took over after Rose and Mel's retirement] to figure out what to do with it. So they set up a fund called the Clean Water Fund and [formed] a committee. [Its focus concerns] first, direct service for people who really needed water immediately, especially people with kids, and those who couldn't afford to buy water and buy groceries . . . ; second, advocacy; and third, education and research." Several other organizations received portions of the funds in their efforts to help those in the greatest of need. One of those was Charleston's Covenant House.

––––––––––––

Ellen Allen (introduced in chapter 4) is the executive director of Covenant House, founded in 1981 to help those with the fewest resources meet their most basic needs: food (including access to clean water), clothes, and shelter. Like many organizations, Covenant House worked tirelessly during and after the crisis to provide clean water and other basic resources. But Ellen's organization went one step further. Fueled by her own physical and emotional encounter with the contaminated water and broader dimensions of the water crisis, Ellen, together with her board, felt compelled to dramatically escalate their concern for their constituencies' precarious situation: "During the water contamination, [we] sued DEP and DHHR [Department of Health and Human Resources]."[11] It was an unpopular stance to take at the time. Indeed, "it wasn't a very attractive proposition to other nonprofits," Ellen explained. "We stood alone . . . We sued them not for financial, not for any monetary gain, but to enforce existing regulations that guarantee clean drinking water." This legal foray was no small step for Covenant House; it risked offending DHHR, a significant funder.

The legal action eventually took the form of a writ of mandamus filed with the West Virginia Supreme Court, which would direct the DEP and DHHR to enforce current standards of the Safe Drinking Water Act. That two West Virginia Supreme Court justices wanted to hear the case Ellen found "encouraging." The other three said it was premature because of pending legislation meant to address the crisis.

Ellen's focus, and that of Covenant House, is the community they serve. "What happens in a situation like this," she said, is that "the most marginalized populations, the poor, they don't have a voice. They really don't have a choice. . . . We see [our legal suit] as systems change, which is another role [we perform]. It's long, and it's arduous, and sometimes you don't know that you're making an impact. But I'm confident that's the only way to go about systems change."

## Activism, Part III

While some may note the problem of complacency in parts of West Virginia (see chapter 7, for example), it stands, insofar as it exists, in stark contrast to the time-honored tradition of activism also found here. Take, for example, the experience of long-term civil rights activist, attorney, and ASWS legal counsel Paul Sheridan.

Paul moved to the Mountain State as an AmeriCorps VISTA volunteer in 1978. He was, he said, summoned here by the Williamson flood of 1977, when up to fifteen inches of rain caused record flooding, twenty-two deaths, and more than $400 million in property damage in the Appalachian states of West Virginia, Virginia, Kentucky, and Tennessee. He dove into environmental work, helping explore the links between flooding events and mining practices. He also met lawyers doing legal aid work whose dedication inspired him to attend law school in Morgantown, West Virginia. Three years later, law degree in hand and a member of the bar, Paul returned to the coal fields of southern West

Virginia and dedicated himself to legal aid for six years. Then he moved with his young family to Charleston and began a career of more than twenty years with the state's Civil Rights Division.

And this brings us to Paul's kitchen, where I interviewed him.[12] He was at the sink, about to drink a glass of water, when he was struck by déjà vu: he was standing precisely here when he drank a big glass of tap water, somewhat discolored, on the first day of the crisis. He remembered the discoloration more than any chemical taste or licorice odor. In hindsight, he wondered if the MCHM contaminant might have scrubbed his water lines to create the unexpected color; after all, its official use is scouring coal fragments.

At a thrown-together community meeting just days after the water contamination, Paul offered his expertise to folks in the audience wanting to file formal complaints to the Public Service Commission (PSC), the state agency that regulates the water company. In the following days and weeks, about ten novice and veteran activists began meeting twice monthly in the Asbury United Methodist Church, whose guiding principle for sharing their building is, appropriately, "Sacred space for social justice."[13] It soon became clear that having Paul with his legal acumen on board was a special asset for the group that would also facilitate and broaden its impact, from providing a sounding board for complaints, to proposing specific features and policies required to transform the water system into a "safe" one, to a campaign to replace the shareholder-owned water utility with a municipally owned water utility.

Paul's legal credentials and left-of-center credo were a perfect fit as the fledgling group, Advocates for a Safe Water System, took its place in the vanguard of a broader chorus of community voices calling for a general investigation by the PSC into the water utility's role in the crisis. The goal of the probe was to uncover critical steps and missteps taken by of the water company that helped set the stage for the crisis. It was also meant as a way to understand the water company's performance in the chaotic days and weeks following the chemical spill. In Paul's words: "The purpose of this [investigation] should be to figure out what happened. Let's get a really good handle on what happened and what the lessons are, what we can learn from it. And then maybe the PSC can enact all the lessons. Maybe we have to look to the legislature for something, or maybe it's the Bureau of Health that has to enact some of the lessons."

Almost a year after the beginning of the general investigation, itself just emerging from an eight-month bureaucratic hiatus, ASWS was also accepted as a party to the water company's rate increase request. Although protested by the water company, this action aligned with the group's intent to maintain a spotlight on systemic problems in the water system and to continue to advocate for improvement. Here, again, Paul was the official voice of ASWS at the hearing. At the time of this writing, the push for a public water utility to replace the private one is just beginning as ASWS further expands its scope. Undoubtedly Paul's experience and expertise will continue to serve the citizen activist community and help articulate its vision.

Minimizing his already-extensive effort to help the community learn from and progress beyond the crisis, Paul understated: "It's nice to kind of have some opportunities to feel like there's something I contribute to, so that's been an important thing for a variety of reasons. But it's been a wonderful experience. I mean, I'm used to representing individuals that have been fired from jobs, and so it's a very different thing to be representing a group of people who are all, you know, all have their sleeves rolled up. . . . I feel like I'm just one part of an effort where everybody is contributing and it's a collective kind of thing."

## On the Meaning of Community

Exactly what does that "collective" mean? Activist initiatives—individual- or group-based, small or large in scope, long or short in duration—contribute to the matrix of a community's crisis response. They help sculpt the social construct called "community" and reveal something of its strength and resilience. A number of interviewees shared their perspective on community, what animates it and makes it strong and what it looked like during the water crisis. Their comments were not part of a statistically designed survey, nor were efforts made to settle on a textbook definition of "community" before they shared their views. Nevertheless, because "community" can mean so many things to so many people, we sought to delve deeper into what activists, especially those involved in the water crisis, meant by it. Three powerful themes surrounding the crisis aftermath appeared and reappeared in our interviews: conversation, story, and trust.

*Conversation.* Threads of conversation and stories woven from them are the fabric of Dave Mistich's life—and his community. He loves stories, especially gathering others' stories and then understanding, comparing, and sharing them. As a reporter for West Virginia Public Radio, Dave reported extensively on the chemical spill and ongoing water crisis. "What I get to do is talk to people," Dave explains, "and then tell other people what we talked about. But the better part of that is just having a conversation with a person. . . . I think, to me, I enjoy just talking to people."[14]

On January 9 Dave began an extensive and extended conversation about the water system with his community, with government and water company officials, business owners, the man and woman on the street, family members, you name it. "It's a very complicated relationship between communities," Dave explained, "and families and industry. I had a conversation with my brother who lives in Georgia now—he's my twin brother—and I said . . . this is going to be one of those things that squirts up, and everyone is going to have something to say about this, and industry and coal, and how it affects people's lives and their safety. I mean it's clear that this is going to bubble up a conversation about public health or whatever. . . . I find it really interesting . . . these relationships between industry and family and locations and livelihood. To me it's absolutely fascinating."

Dave saw the crisis and community response affect other environmental concerns in the region. In one example, the dynamic of West Virginia's rural and urban communities and their access to clean water entered the public airways briefly through some of West Virginia Public Radio's programming. When I raised the issue of residents of southern West Virginia whose water wells had been fouled for years by mining activity, Dave responded that "our reporter down there, Jessica Lilly, was aware of some of these issues in Wyoming County. . . . When this water crisis happened it gave an opportunity for her to say, 'Things are really bad in Charleston . . . and they're having this water crisis, but, you know, an hour and a half, two hours south, these problems have persisted for years.' The water crisis certainly created an avenue for us to shine light on problems that exist around the state in other situations."

In another instance of collateral awareness, in which concern over one issue transfers to another, Dave said, "I think the crisis absolutely provided a reference to other environmental issues. . . . I think that the situation with the Keystone [mountaintop removal coal strip] mine at Kanawha State Forest [again, within a mile of the Charleston City limits], that whole situation, the momentum that was carried from activists and from environmentalists from the water crisis might've dovetailed into that movement. . . . I think if you talk to some of the people that have been involved with both . . . they could probably draw a line between their interests." This comment also recalls Paul Epstein's experience when he chose to broaden his environmental focus based on his water crisis experience.

When asked about his trust in the water system, Dave reframed the issue as one of curiosity. "I wouldn't necessarily call it a trust thing," he said, "as much as a curiosity thing. . . . It made me a lot more curious. . . . I was not just answering questions for the public, . . . that's not just what journalists were doing and reporters were doing here. But they were also answering questions for themselves because they are that public. So I think that . . . questions of trust or curiosity . . . were answered at some point in the process of tackling this story as a collective group of Charleston-area media. . . . It made me think, too, about the difference between a mindset that wants to blame and one that wants to find out, you know what I mean?"

After hearing and talking about so many perspectives on the crisis, Dave had something to say about its impact on the larger community. "It was a really complicated thing that happened," he pointed out, "and I think that a lot of this is much broader than a chemical spill and three hundred thousand people's water. I think that events such as this . . . are metaphors in a sense; they're living metaphors. They're events that explain wider things that need to be talked about. . . . I don't know that we've figured out a way to distill it down and to get it to a point to where everyone can agree on where we need to go next."

*Story.* If Dave explores the role of conversation in understanding connections of family, industry, and community and how these were impacted by the crisis, then Paul Sheridan articulates the narrative structure framing them. First is the event itself; something, typically many things, happened in a historical and physical context usually less than obvious. What follows is the thought, conversation, and curiosity about the event that construct its narrative. Those who create these stories have different experiences, beliefs, and motivations that give rise to altogether different perspectives even about the same event. These stories play pivotal roles when they are used to assign blame or reinforce cultural norms, describe how the crisis unfolded, or assess its impact on the community.

All of this may seem obvious. But from the punishing poverty in southern West Virginia's coal enclaves, where he spent almost ten years to his more than twenty of litigating civil rights abuses in Charleston, Paul Sheridan became intimately familiar with the power of narrative, especially its impact on communities in crisis. In much of Paul's experience, regardless of its source or motivation, the story was tragic from start to end; often there were no happily ever afters. Sometimes, though, he saw communities use their crisis as a catalyst to script, if not a Hollywood-style feel-good finale, then something more modest and believable, perhaps something simply positive, as the curtain dropped.

"One of the things that I've done in the area of civil rights over a number of years," Paul pointed out, "is work with some people who [work] nationally, . . . documenting communities that stand up against hate and intolerance, [often] filmmakers, and they typically have gone to places where there's been some kind of awful event. . . . And they're about telling the story about how the community's responded. And working with them has helped me see that the question is whether the narrative you focus on is . . . the disaster or whether it's . . . on the response to the disaster. It's really a kind of consciousness thing; it's really what you pay attention to."[15]

"There are some communities," Paul continued, "that don't ever get their act together to respond to the awful things that happened to them, so the story really remains about the disaster. But some communities, they really rewrite their history. The story that becomes their story is . . . how they responded. And it's always seemed to me that this is that kind of opportunity for this Valley. We could be looking back from ten years out at the community that not only came through this but used it as a . . . catalyst to address its environmental problems, its health, whether we're talking air quality or water quality or drinking water safety, you know, the regulatory environment, all the kinds of things I think have really been an ongoing problem. This could be the moment when we really put that into balance. And also the economic development things. . . . It's the same kind of community energy that makes an economy go if people come together and say, 'This is my home and not only do I want a safe water system, but I want a school system that I'm confident in, that I feel gives my children a future. And I want an economy that's going to be serving the interests of people, that's going to be stable, that's got long-term potential.' So I think all of these [issues] are related."

In many ways, Paul encourages a reframing, in which a crisis cudgel becomes a crisis tool, something to be used instead of just endured. The water crisis, in particular, created its own spectrum of new circumstances and stories about our community. In this process, traditional narratives were altered or shattered altogether. Narratives surrounding trust and safety are examples. "All of sudden," Paul said, "not only the convenience [of safe tap water] can't be counted on for a period of time, but far more significant than that, the safety kinds of things and the feeling that . . . somebody must've been looking into this, right? Somebody had planned for this and had contingencies thought through? The idea that the stuff could get into the water system and probably into a glass of water without anybody even knowing it . . . [not even the water company]. . . . [It's] now my strong suspicion that the system was entirely contaminated before they were even alert to what was going on. . . . [This] is appalling to me. I thought, 'This isn't the world I thought I lived in.' I thought I lived in a world of government overregulators who are hovering around making sure that everything is more than alright and the great threat is to our freedom and our independence . . . from these government overregulators. And to find out that nobody had looked at the

[MCHM chemical] tank for more than twenty years . . . [that there had been] no systematic detection [or continuous chemical water monitoring] at the front end of the water system."

Paul suggested that these are the stories that really matter, that shift our ways of thinking about things like water. "So that stuff has been far more [alarming] than a week of bathing out of a bucket," he argued. That's "been the thing that's affected me. . . . I've been fighting different battles for many years and I don't think of myself as somebody with great illusions about either the goodness of the corporations that bring us the goods and services we consume or the consistent thoroughness of all government investigators, many of whom I've worked alongside of in different ways. This was like a new level of shock and disappointment."

As Paul suggests, we build stories about how our world works, which helps us frame and negotiate our daily routine with a degree of poise and assurance. When those frames are broken, we have to build narratives we feel are closer to the "truth." Paul summarized: "There are all of these different stories. So one of the [questions] is, 'Which ones will become the defining ones? Which ones will gain traction?'"

*Trust* (or *What will help restore our trust?*). The sudden realization that our water could be contaminated without our knowing it—and, worse, without the water company knowing it—shows just how invisible some of our social contracts are, and how fragile. Once trust has been broken, rebuilding it may become as important as mitigating physical consequences of the crisis itself. Aside from a return to normality and the healing of trauma afforded by the mere passage of time, just how will people once again come to trust that the water they drink is safe? Given our history in West Virginia and especially here in Chemical Valley, just telling us "it's safe" won't be enough.

Surrounded by the ebb and flow of clarity and confusion, several interviewees shared their views on the importance of community knowledge and ongoing involvement to transform the water system and restore their trust in it. Many of them, indeed, many of us writing this book, continued to distrust the water long after the water company or the government said it was safe. For example, when we interviewed her eight months after the crisis, Reginia Lipscomb, a retired community development banker, reported that "to this day I still don't drink or cook with [tap water]. I still buy water. I just have a hard time trusting some of the politics around here. . . . You would think they would do an assessment [of chemical storage tanks near the Elk River] and be able to report out, 'OK, we've combed the whole Elk River and these are the companies. . . . There are some hazardous chemicals stored here, here, and here.

We need to get these moved.' The public [has] not been advised what else is out there. I think we need to know. . . . You and I could be still sitting here five years from now, and they're still 'working on it.'"[16]

Linda Koval, who works as a tutor at West Virginia State University, shares similar concerns: "I feel like I'm more at risk. There are not really adequate procedures and regulations in place to necessarily prevent this from happening again. I know that the legislature did take action to try to regulate it better; that's only going to work if people within government and within private industry have the will to implement and to maintain a certain level of due diligence past the crisis. Once people's memory starts fading, you know, industry and government may think, 'Oh, we're safe now. We don't have to really abide by the letter or the spirit of the law.' I think we need continual public attention to this and to elect people to government who are not beholden to the chemical industry or the coal industry, who really have the welfare of the general public at heart. . . . I know you're never going to get rid of the influence of the chemical industry or the coal industry. But I think if people are just better informed and, as a result of this crisis, that there is some responsibility there for individuals, that that may help. If people are just alert and not so complacent and assume the government may or may not be doing or that industry may or may not be doing [something]. . . . [There should be] more awareness and more attention to this whole issue of water quality. . . . That's a really good thing to take a hard look at this and to have so much attention focused on it. . . . Probably, there are enough people in the valley concerned with the quality of the water that they're going to be looking at this and asking questions and not let it go, which is what I hope will happen."[17]

David Chairez echoes the need for citizen participation and underscores the importance that it be community-wide and persistent. "How can I really trust that everything is OK, that I'm going to be OK, that my kids are going to be OK, if I continue to use and consume the water? . . . For me, personally, it's a matter of time. I think it's a matter of consistency, and I think it's an acknowledgment that, number 1, we really don't know; number 2, we will continue to be mindful of this and, really, it just has to be a consistency of behavior, that, if an issue arises, we do acknowledge it, we do address it, we do attempt to work through it together as a community and people begin to see that consistent action and feel like, 'OK, we really are addressing this together and it wasn't just a sound bite.' . . . The most important thing I'm encouraged about is that, through this process of citizen-led dialogue and deliberation, awareness of the water system's ongoing vulnerability to chemical spills hasn't just gone away. I'm encouraged that things are being done, that people are continuing to take this seriously and that there are people who say, 'We didn't just survive this. Now we've got to come together and get ready for whatever is to come.'"[18]

## Notes

1.   David Chairez, interview with Jim Hatfield, August 5, 2014; with revisions, personal communication with Jim Hatfield, July 25, 2017.
2.   Shelda Martin, interview with Marla Griffith, August 26, 2014.
3.   Marlene Price, interview with Trish Hatfield, September 11, 2014.
4.   Lucy Nelson, interview with Marla Griffith, July 15, 2014; with revisions, personal communication with Marla Griffith, August 3, 2017.
5.   Belinda Vance, interview with Cat Pleska, July 9, 2014.
6.   Becky Park, interview with Trish Hatfield, September 10, 2014.
7.   Austin Susman, interview with Marla Griffith, August 14, 2014; with revisions, personal communication with Marla Griffith, August 3, 2017.
8.   Paul Epstein, interview with Jim Hatfield, September 9, 2014.
9.   Martha A. Ballman, interview with Jim Hatfield, August 26, 2014.
10.  Rose Edington, interview with Jim Hatfield, September 4, 2014. Rose also pointed out that the women brought water to the gathering to symbolize they were coming from all four directions of the land. The ceremonial mingling of the waters symbolized the mingling of diverse ideas and places to create something new for the faith.
11.  Ellen Allen, interview with Trish Hatfield, September 10, 2014.
12.  Paul Sheridan, interview with Jim Hatfield, August 26, 2014.
13.  Joe Jarrett (pastor), Asbury United Methodist Church, Charleston, personal communication with Jim Hatfield, June 12, 2014.
14.  Dave Mistich, interview with Jim Hatfield, September 13, 2014.
15.  Sheridan, interview with Hatfield, August 26, 2014; with revisions, personal communication with Jim Hatfield, August 23, 2017.
16.  Reginia S. Lipscomb, interview with Jim Hatfield, September 10, 2014.
17.  Linda Koval, interview with Trish Hatfield, August 12, 2014.
18.  Chairez, interview with Jim Hatfield, August 5, 2014; with revisions, personal communication with Jim Hatfield, July 25, 2017.

# CHAPTER 9

# WVWaterHistory.com and Producing Digital Resources on a Water Crisis

Gabe Schwartzman

This oral history and collaborative ethnographic project has shown how West Virginians reacted to the MCHM spill of January 2014 and how this spill affected their lives, political consciousness, and sense of safety. Perhaps most importantly, this research has worked to place West Virginia's chemical spill within the context of other disasters (as, e.g., in Hoey's chapter 2). We see this West Virginia chemical spill not as a stand-alone event, as much of the new media and many politicians would have us believe, but as a contingent part of the world we live in.

As the authors have explicitly and repeatedly pointed out, in the extraction of natural resources, the flow of people and things to and from one place to another, and in the midst of our daily lives, disasters are neither natural nor unusual. They are mundane—fundamentally part of how our society functions. Furthermore, the disaster is not simply that a chemical spills, an oilrig explodes, or that water stops flowing to citizens. The disaster is in the human and ecological dimensions.

In this chapter I briefly revisit the human elements of disaster and continue in the vein that disaster is not natural but human-made. The disastrous elements are the impacts that people suffer, and the great majority of suffering during disasters is attributable to systemic inequality. As was the case in West Virginia's chemical spill, the spill was a disaster for those most vulnerable, while an annoyance, if a frightening one, for many other West Virginians impacted during the spill.

## Building a Website

I came to this project after it was already well underway. I was working on a website at the time, which came to be called WVWaterHistory.com. Working with web developer and graphic artist Alicia Willet, I assembled the website on a post-undergraduate grant from the University of California, Berkeley. I had arrived in West Virginia before the January 2014 spill, intending to research stories of water crises throughout Appalachia. The spill clearly focused my studies on that one incident.

As the water crisis unfolded, I decided to explore how people living with other water contamination issues in the coal mining areas were coping with this new spill and investigate their reactions to the new spill. I chose to interview advocates and leaders in the affected communities, including some from marginalized communities, whose stories spoke to the systemic and widespread nature of water crises in the mountains. I presented my findings on the website to make the research available for policy advocacy groups and to bring participants' interviews together with graphics, maps, and photos, melding these media in one interactive space.

I met Eric Lassiter, and then Beth Campbell while conducting these interviews. It was under their direction that I became familiar with the idea of a collaborative ethnographic process. As a result of our conversations, I decided to take all the work I had planned to post on the website back to interview participants and ask for their input. At first I was hesitant. What if a participant did not like what they had said or how they had been represented? Would allowing participants to influence their own representations undermine my research? While the process initiated further discussion and interpretation, my concerns quickly dissipated as I realized how the method could strengthen my content, as well as my relationship with the research participants who had so freely shared their experiences with me.

For the purposes of the website, I decided to organize the interviews thematically and to highlight two major themes that surfaced in the interviews. The first theme had to do with the inequality of the spill's impacts. For a variety of reasons people experience disasters in very different ways. Many of the people I talked to recognized this and often pointed it out. In contrast to how the media often represents such disasters—as momentary experiences of suffering—I found some of those most vulnerable were well aware that these disasters were neither momentary nor accidental. The website sought to tell stories in such a way that recognized and acknowledged these discrepancies.

An important theme that emerged in the interviews along these lines concerned vulnerability. Marginalized people—people in poverty, people of color, incarcerated people, and rural people—were disproportionately affected during this crisis, of course. Race is an illustrative example, clearly important during the crisis, because of the way that some of the poorest historically African American communities were served during the emergency response.

Two of the interviewees featured on the website—namely, Reverend Matthew Watts and Crystal Good—speak directly to this issue.

A second theme I explored on the website had to do with the story of disparate vulnerability during the chemical spill as experienced by rural people living in the southern coalfields of West Virginia. Not only were many of these communities served poorly during the emergency water response; they have also lived with water contamination in one way or another for many years. In some of these areas, coal mining has been sullying well water and streams for the better part of a century. Rural people who had consumed mining contamination before experienced the chemical spill quite differently from those in Charleston. This was especially true for those people who had recently been brought into the city water system as a means of compensating for their contaminated well water—in Prenter, West Virginia.

Prenter provides a potent example of how disasters can be nothing new for some people. Massey Energy, an infamous coal company, now defunct, had pumped coal slurry (the liquid residue produced by washing raw coal) into underground mine shafts for years, and the slurry eventually flowed into the area's groundwater. After a series of cancer studies, the community sued the coal company, ending with a large settlement, closing off Prenter's wells, and connecting its homes and businesses into the city water system. Several of the people I interviewed from Prenter are featured on the website, including Maria Lambert and Daniel James Estep, both of whom spoke to the despair and sense of hopelessness after the January 2014 chemical spill. They had believed the promises that connecting to city water would mean water security. It did not, of course.

For me, exploring these two themes brought to life the idea of disaster presented throughout this book: that the impacts of disasters are not accidental, abnormal, or "natural" but rather very much based in human decisions and policies. My participants' lived experience of the unequal and continual disasters displays an astute consciousness of the nonaccidental nature of natural disaster.

## Interviews and Maps as a Digital Storytelling Platform

For the website, I took the interview content and combined it with historical records and data about how drinking water systems evolved in West Virginia. I assembled the website not only for academic use but for use as a resource, mixing data and analytics about the water company, the water system, and the chemical spill with the stories of those who experienced it.

The history I analyzed on the website included the mechanics of the water system, the legacy of public water systems in rural areas, and the expansion of private water across the region. The website also required visualizing distributions of inequality, during both the spill and ongoing water crises. I worked with graphic designer Alicia Willet to make maps of this inequity and of the legacy of spills in the area. Importantly, I chose images, maps, and other graphics that would be useful as educational tools, particularly for activists. I also chose to develop graphics that could be used in social media and images that were compelling as part of an advocacy campaign. To that end I also used parts of the interviews I thought were most compelling for advocacy reasons on the website.

This decision to focus on advocacy was ultimately my own, but the collaborative ethnographic process reaffirmed that decision. As I neared the completion of the website, I began to return my representations of people's work and interviews to them for comment and discussion before publication. All my participants originally discussed the need for advocacy in our interviews and agreed that their words and their ideas could be used to advocate for change on the website. When reviewing their materials for inclusion on the website, many agreed on the advocacy angle. Some of these collaborations, though, were not as cut-and-dried as simply talking with the people whose ideas I used. In one instance, for example, one of my consultants decided she was uncertain about her involvement in the process. Though at one time she thought of herself as an activist, she hesitated to be represented as an activist again. But upon further discussion, we found a middle ground, making adjustments, for instance, to her images in such a way that eased her feelings about being presented on the website.

## Beyond the Website

Since I published the website, it has now become a resource for activists and policy advocates building a case for a public water system. The maps, graphics, and content are all publicly available on the website. They are consistently used by the organization Advocates for a Safe Water System in the PowerPoint it presents on the need for a public water system. Voices of Appalachia, another advocacy organization, used the data, photos, and graphics in a central article on water privatization. The website and the ethnographic content now has a life outside of its academic contexts.

In the end, the ways that we as academics, advocates, and citizens talk about and think about disasters is critically important. As the authors in this book discuss, it is quite easy to see the West Virginia chemical spill of January 2014 as the inevitable consequence of living in an extraction state, in rural Appalachia, where environmental problems seem constant. But as authors in this book discuss, there is power in seeing the event as part of a larger system of disasters, as well as seeing the way that vulnerability is unequally and consistently distributed in southern West Virginia. Seeing the spill in this framework sheds light on the policies and political actions that could serve to reduce vulnerability for those most at risk.

# CHAPTER 10

# What Does a Water Crisis Sound Like?

Laura Harbert Allen

When I was a child, my brother, sister, and I played a game every time we came to visit family in Charleston, West Virginia: Who could hold their breath the longest? Our goal was to breathe as little as possible as my dad piloted our big green Oldsmobile 88 through I-64 from Charleston to Huntington. We played this game because driving through the Chemical Valley in the 1970s was a smelly enterprise. The air was heavy and crushing; I remember complaining to my mom that it was like rotten eggs or the worst, most medicinal cough syrup imaginable. I was always relieved to get to the city of Cross Lanes, where the smell finally began to recede.

My siblings and our childhood game were on my mind as I walked home from work on January 9, 2014, at about 5:15 p.m. At the time, I was director of communications for the United Methodist church in West Virginia and on deadline for our monthly print publication. I had taken my dog, Miles, with me to the office, which was walking distance from my home on Charleston's East End.

I suddenly realized that the air smelled strange. Worse than usual. As you have no doubt realized in the stories throughout this book, including my childhood experience shared here, strange odors in Charleston are normal. But something was different this time. I looked down at my dog and realized he was breathing heavily. As we made our way along Washington Street, I grew increasingly worried, so much so that when I got home the first thing I did was go to Twitter, and sure enough, something was wrong. My feed was full of warnings to avoid touching the tap water. Both West Virginia Public Broadcasting and the *Charleston Gazette* announced that the governor would hold a press conference at 6 p.m.

At that point, I texted my colleagues, telling them not to touch the water and alerting them to the governor's press conference. I called the bishop of the West Virginia Conference, and we began drafting a statement and planning the response of the United Methodist church in the nine counties impacted by the spill.

By this time, it was about 5:45 p.m. I jumped in the car and headed to our neighborhood Rite Aid and bought three cases and five gallons of bottled water. "Have you heard about the chemical spill?" I said to the clerk as I checked out. "No," she replied. "I think you all will be pretty busy tonight," I said. That was the last time I bought water in Charleston for a week.

I was lucky. Many friends and neighbors, including Eric Lassiter and Beth Campbell, drove for miles to find bottled water.

The days that followed are a blur. I spent them out in local churches around the Kanawha Valley, documenting which ones could provide help to local residents. Saint Andrews United Methodist Church in Saint Albans, West Virginia, invited anyone to come and fill up whatever containers they could find with their tap water. Parts of Saint Albans were served by a public service district that was not impacted by the spill; thus the church's water was safe to use. Tanker trucks filled with water made their way to Charleston and at least one of them set up at Aldersgate United Methodist Church in Sissonville, a community just outside of Charleston. As the week wore on, the community got back on its feet. But a distrust of the water lingers three and a half years after the spill. Grocery store shelves are often bare of bottled water.

The water crisis was a personal and professional turning point for me. I decided that I wanted to go back into making stories in the public radio tradition. Within a month of the initial crisis, I started a class at the Center for Documentary Studies at Duke University. As I drove back and forth from Charleston to Durham each week, an idea for an audio documentary began to emerge. By early summer of 2014, I began what we in audio and radio making call "gathering tape." I began talking with people I knew in town, including a woman who cleaned houses for a living named Angela Walker. I pitched my story idea around and was commissioned to make a seven- to nine-minute documentary-style piece for *Making Contact*, a public radio social justice–oriented show produced in Los Angeles that aired on about 130 stations around the country at the time. The piece aired as part of an episode entitled, "Not a Drop to Drink." (That story, which follows the episode's introduction, can be accessed at http://www.radioproject.org/2015/08 /not-a-drop-to-drink-our-dwindling-access-to-clean-drinking-water/.)

Angela represented a group of people overlooked in the mainstream media's coverage of the water crisis. She was a barely middle-class, blue-collar, independent businesswoman who relied on water to make a living. She spent

January 9 at a client's house cleaning and "had her hands in the water all day long." Angela's story is important because it showed the economic impact the water crisis had on a family that already lived in the jagged boundary between blue-collar working class and working poor. Her husband was an independent housing contractor, and he and Angela lost income for months after the crisis. Cleaning clients canceled because they did not want their floors mopped or dishes washed with tainted water. They lost an expensive power washing tool because customers did not want tools tainted by the water used in or around their homes.

The couple had three children to support, and the economic losses the family suffered sent them into poverty. "I find myself in line at food pantries and on food stamps," Angela said. "We needed clothing vouchers for the first time. The year before [the crisis], we bought clothes for some families. . . . It's just totally shifted."

The impact of the water crisis was ongoing, I realized, and so I decided that a larger project was a logical next step. I began collecting interviews with friends and neighbors. Through the course of interviews with my fellow Charlestonians, I was introduced to Beth Campbell by my good friend Rebecca Roth (see Beth's interview with Rebecca in the introduction). As Beth and I talked, it became clear to me that there was a good possibility for a partnership between my work and the oral histories Beth and others were gathering with her husband and colleague, Eric Lassiter, who directed the graduate humanities program at Marshall University.

The potluck dinner Eric and Beth hosted at their house for the project collaborators was indeed serendipitous. It became clear to me there that I had found a home, a place to explore the intersection of journalism, documentary, and ethnography. More than that, I'd found a community of people who were intellectually curious and active from a variety of disciplines.

I agreed to become part of a small graduate seminar class at Marshall University's Graduate College, which, as discussed in chapter 3, began the process of interpreting and writing up the oral histories as ethnography. Gabe Schwartzman joined us to talk about his post baccalaureate research on the integrity of the water supply in southern West Virginia (see chapter 11). And I came on board to lead an effort to make an audio documentary with the class.

My principles as a storyteller and documentarian, learned through my years in public radio and in my studies at the Center for Documentary Studies at Duke, would be pushed to the limit through the semester. Who was the character in this story? How would we build action, create a beginning, middle, and an end? Who would get to make final editing decisions? I was concerned that, as interesting as our conversations might be, making a compelling story out of a graduate school seminar, along with the oral histories from the larger team's oral history research, would be challenging. I could not think of a single place I could pitch this kind of piece to a public radio news director.

I soon learned that was not the point.

Using Jessica Abel and Ira Glass's *Radio: An Illustrated Guide* as one of our texts, we started our work making personal stories. My goal was simple: to get students comfortable with audio and to learn by doing. I asked them to write a short essay about a childhood memory that included water of some kind. It could be a dash through a rainstorm, a creek where they played as children—anything at all. We talked about the difference between writing for print and writing for the ear. Radio is perhaps the most didactic medium there is. I told them to describe in detail their scene through their senses. What did the water look, taste, smell, or feel like? I also instructed them to describe the details of any object in their memory, to bring to life a memory in a way that the rest of us could experience.

The results were powerful. To hear some of the results, see the section "A Shed, a Boat, and a Water Project" on Source Material: West Virginia Conundrum (https://watermarkwv.wordpress.com). One student, Joshua Mills (coauthor of chapter 6), talked about his family farm, which sits directly across from the Cabot Pumping Station (an artifact of hydraulic fracturing) in Wayne County, West Virginia. The *whim-whomp-whim-whomp* of the pump and the gurgling of a creek both cut through the aural landscape of the place where he grew up.

Josh also talked about building sailboats with his grandfather with wood harvested from the family's property. I was intrigued by the tension inherent in Josh's experience. He grew up on land that was naturally beautiful yet tainted by industry. I asked if I could come out for a visit and see—and hear—this juxtaposition for myself.

Around the family's kitchen table, between bites of spice cake washed down with coffee, Josh told me that this family land was special: "I probably became an archaeologist because of my childhood here,'" he said. He described going up the hollow, jumping across the creek, spearing crawdads, a young Indiana Jones seeking adventure.

Josh took me on a tour of his grandparents' property. We stood by the creek, and through my headphones I heard the tumbling of the water over rock. In the background the *whum-wump-whum-wump* of the pumping station cut through the rock, creating a feeling of insidious violence. "I just wish it would go away," said Josh.

We spent some time in the work shed, about twenty-five yards away and slightly uphill from the main house. An old wooden sailboat hull was set up on a pair of wooden sawhorses, a mottled gray-blue-brown husk that told of years of use. Josh meticulously applied polyurethane along the seams of the boat as his grandfather looked on.

Josh told me that the graduate seminar made him see things differently. He wasn't in the spill zone during the water crisis and admitted that he didn't really feel the impact of what happened right away. "I felt sorry for people, but I didn't think about it too much," he said. By the end of the semester though, Josh's awareness of water had deepened. The Elk River chemical spill and the ensuing water crisis tainted his memories. The creek on the family farm and the waterways of West Virginia where his boat sailed were no longer pure.

This fall from grace and purity in memory was a common theme. I can still see the childhood bike of another seminar student, Emily Mayes (author of chapter 7), in my mind's eye. It was white and pink with tassels on the handlebars, ridden regularly up and down a road where her mother and grandmother lived. The road was also next to a creek in Kanawha County that sometimes flooded. "Was that water safe? I don't know," she said during one class session.

Another student, Jay Thomas (author of chapter 5), shared his experiences at a cabin in Pocahontas County, West Virginia, as well as his experience of running a restaurant during the spill (which he describes in chapter 5). I can see his family gathered around a table while the family matriarch fries fish caught in the creek that morning.

Hearing these stories triggered my own memories of water in Appalachia, and a theme began to emerge through the class: the juxtaposition of the stunning natural beauty of West Virginia—"like living in Switzerland," as Jay once put it—with the pollution that has occurred here so that industry can profit. How do we as West Virginians reconcile this Dickensian extreme: to live simultaneously in an extremely beautiful place and an extremely ugly place?

Throughout our work during the semester, we read two other books: Lassiter's *Chicago Guide to Collaborative Ethnography*, to get a sense of the collaborative framework for the project, and Timothy Button's *Disaster Culture*, to help situate our water crisis within larger contexts. These books, along with what emerged through classroom discussions, were important to me as an audio producer—and, indeed, a human being—for a couple of reasons. First, I recognized anew that we as Appalachian people were not alone. As Button makes clear, the water crisis and the way it was handled by government and industry officials followed a familiar script that was both infuriating and galvanizing. What happened in Charleston, West Virginia, has happened before:

in New Orleans; in Prince William Sound, Alaska; and in the Gulf of Mexico. The circumstances around Hurricane Katrina, the *Exxon Valdez* spill, and the *Deepwater Horizon* explosion are different, but the scripts are eerily routine. It is difficult to fight back against corporate interests, especially in a place like West Virginia, where there is a long history of the conflation of those interests with government. Learning about what happened in other places informed the work we were making with voices and sound. And it helped me personally to recognize what had happened in the place I call home.

The other insight I gleaned during this time came from Lassiter's book and the work that we all did together to listen, voice, and convey through audio what it meant to experience the water crisis. Eric outlines some of the methods of collaborative ethnography in chapter 1, but importantly, in using collaborative ethnographic methods to shape audio documentary, it was important for us to share and negotiate editorial decisions, in this case, with me, the producer and editor, as the content emerged. Such a process puts up front the power relations between the person (or persons) who comes into a community to make content, and the people who help her create it. The labels of "subject" and "informant" are the traditional terms used by ethnographers and documentarians. Doing audio production like this meant using words like *partner, collaborator,* and *consultant* when referencing my work with people in the making of a documentary. But, I want to make clear, these are not just labels: they signify and describe a different kind of relationship, in which partners are as involved in the making and editing of the final documentary as is the editor or producer.

This is important. Producing collaborative work like this can be difficult and challenging. Progress is slow because there are different opinions and experiences in the approach to the work. It is different from producing conventional documentary work. Any form of public media content is, of course, somewhat collaborative. The conventional process goes something like this: a producer pitches a story and works with a news director or editor to make a piece. A good relationship with an editor is one of the best forms of collaboration that exists. For me, a good editor sharpens and suggests ways of improving the work made by a producer. They check the natural impulse to make a producer tell a story the way they would. A good editor-maker relationship is defined by trust. In large projects—such as when I produced stories for public radio—that trust can extend to a host of others, like radio engineers, announcers, sound editors, executive producers, and so forth. But that trust exists within the professional boundaries of media-making relationships. In other words, I trust the engineers and announcers to do the jobs they have been trained to do. They are getting paid to do the work, so I assume it will be done. I have no say whatsoever in how engineers do their job. Ultimately, when an announcer turns on their mic to go on the air, I have no real control over their vocal tone or whether or not they will butcher my introductory copy. Conversely, they have no input or control over what I make. The project at Marshall was different because we were comakers who all had a say in the final product.

That, of course, also opens the issue of vulnerability. In conventional audio production, sharing creative work can, of course, feel pretty vulnerable to begin with. So the idea of opening up the process further to still others—the "subjects" of the documentary—quite frankly made me nervous. It was completely uncharted territory for me, and it took me a while to get comfortable with it. My biggest concern was one of scope: when everybody has editorial control, how would we ever reach agreement? How would we make a piece that people would want to listen to during local news cutaways in NPR program such as *All Things Considered* and *Morning Edition*? Long-form documentary work is not something traditional NPR news programming has aired in a long time.

The advent of podcasting, though, has revolutionized the way audio is heard and distributed in America, and, despite what some might think, there is still a rigorous process when it comes to getting work heard on podcasts. For example, editors don't like stories that are already produced because they want to be part of the process of creating the work. They know what their audience wants (in theory at least). In fact, part of pitching for a savvy producer is convincing an editor that your story idea is one in which their audience would be interested. This stands in pretty stark contrast to the ethics of collaborative ethnography, which are not concerned with editing for the same kinds of audiences.

What emerged through the semester was an in-depth look into the water crisis and Appalachian identity. We called my part of the work "collaborative documentary," and I suppose that is the best description for it. It is nonfiction, factual recording of people's reaction to an environmental crisis. This project, however, also infused artistic components into the work that are not always associated with the American audio-making tradition. And yet the audio produced was a form of media art that mixed snippets of story with ambient sound and music. It was a hybrid form of audio making that was not traditional public radio. It had elements of documentary and ethnography but was not rooted exclusively in either discipline. As of this writing, the work is still emergent, but I'm confident that the collaborative piece that winds up getting made will reflect a shared vision. It just takes a little time. For an early sample of the documentary, which pulls together student work and the oral histories, see https://soundcloud.com/laura-harbert-allen/ive-been-down-so-gd-long.

As I write this, from a documentarian's point of view, the collaborative style of the work doesn't nicely fit into a single twenty- or thirty-minute story. Ultimately, I think this work is best suited for some kind of interactive, transmedia website. Put simply, *transmedia* means that various forms of media are presented to tell a story. Audio, video, writing, and still photography are combined to create an in-depth portrait of an event or people and place. (With this in mind, I've posted much of this work in its various forms on https://watermarkwv.wordpress.com and https://soundcloud.com/laura-harbert-allen.) I also believe some sort of long-form, ongoing community storytelling component would serve this project well. For example, a way for those who suffered through the crisis to record how they feel five, ten, or twenty years after the chemical spill. The internet makes this kind of long-term in-depth ethnography more doable than ever before. I envision people recording their thoughts via smartphone and uploading them to a site of some kind. More traditional methods could also be used—an interviewer could do follow-up interviews, with the subjects of Lassiter's original oral history project. Those pieces could also be uploaded online and backed up on a hard drive for archival purposes.

In any case, it's natural through the course of projects like this one to look back and think about what could be done differently. This work requires an ability to live with ambiguity. Themes and methods emerge, and conflict is inevitable. The key thing to remember is that healthy conflict makes good work. Navigating the confusion requires patience, but it is worth it.

If I were to give someone advice who wanted to do a project similar to this one, I would tell them to be prepared to feel confused. Lean into the tension and conflict of the unknown, and be willing to try things that are not comfortable. I would also say that ground rules are important. What are your goals? What are the group's goals? How are those outcomes decided? These questions were difficult to answer at times. The biggest challenge was not content—plenty of that was made by the group. The biggest challenges are how to make the content cohesive while remaining true to its collaborative spirit.

Big collaborative projects take time, and this one was no exception. The project never had traditional public media news or documentary constraints. Because this was a project about people and their response to a crisis, the response is still ongoing. It never really ends. And so, by definition, sharing their stories never really ends either.

# CHAPTER 11

# Can We Trust the Water System Now? Some Updates

Jim Hatfield

More than five years after a chemical spill expanded into a regional water crisis, can we trust the public water system? In the chapters of this book, this issue surfaced time and again. That the public's trust had suddenly been broken was signaled by the governor's "do not use" order. As we have seen in this book, at least one interviewee continued to use bottled water for drinking and cooking more than three years after the crisis, and another credits the breach as a key factor in his and his wife's decision to sell their business and leave the Kanawha Valley.[1]

A brief summary of the crisis trajectory helps answer this question and appropriately begins with the decrepit tank whose contents smelled like licorice. Freedom Industries' six top officials were found guilty of criminal violations, fined, and sentenced either to short jail or probationary terms.[2] The tank farm has since been reduced to scrap metal, and the site itself was auctioned off in December 2018 and may be used by its new owner for equipment manufacture.[3] This transformation is an unqualified improvement to the water system, even though the financial and emotional price tag for the entire region was unnecessarily high.

Less clear is the impact of low-level ongoing chemical drainage from contaminated soil at the site, though some soil has been removed and other steps taken to minimize runoff. It is still difficult for citizens to learn details of sampling frequency, procedures, and results from WVAW for MCHM sampling at their intake.[4] Of continuing concern, too, is the activity of former Freedom executives. Do they, for example, now own or manage other chemical storage sites, and more important, are those sites located in the vicinity of a water treatment plant?[5]

West Virginia State Code specifies the roles and relationships for state agencies with the water company, WVAW, a corporate subsidiary. One agency, the PSC, determines utility rate structures and creates and enforces rules for West Virginia utilities. It can also initiate general investigations, quasi-judicial hearings, to probe the performance of utilities as it agreed to do six months into the water crisis. The probe eventually spanned thirty-two months due to commissioner recusal and turnover and multiple procedural objections raised by WVAW.[6] It was further complicated because of a concurrent class action suit against WVAW and Eastman Chemical Company playing out in federal court and litigated by some of the same lawyers.[7]

On January 25, 2017, a settlement in the PSC General Investigation was reached: WVAW conceded several plaintiff issues but was allowed to seal expert witness testimony as well as avoid cross-examination and any admission of wrongdoing in their response to the chemical spill. Most of their concessions had long been championed by the nonprofit Advocates for a Safe Water System, which promoted a more robust and contamination-resistant water system. Paul Sheridan (ASWS lawyer and an interviewee met earlier in this book), under the guidance of ASWS, however, did not sign the settlement. He explained that although the commitments made by WVAW were positive, the decision to settle left too much information undisclosed, information that might help avoid future contamination events.[8]

On balance the PSC General Investigation was a qualified success in reducing the water system's vulnerability to chemical spills, something like a baseball record book entry with an asterisk. It nudged the system farther along its path to consistent and reliable operation even in the event of an upstream spill, but could it have been a much bigger push instead?

Overlapping the PSC General Investigation was the United Stated District Court class action suit, *Good et al. v. American Water Works Company, Inc.* (the parent company of WVAW). Like the PSC investigation but sixteen months later (June 2018), it, too, ended in a settlement, only this time serious money was involved: American Water Works Company agreed to pay affected residents and businesses $125 million and Eastman Chemical, the manufacturer of crude MCHM, agreed to $25 million.[9] Like the PSC action, this was another qualified success. The remuneration was significant: although it covered only a fraction of the costs for replenishing potable water, lost employment, and lost business revenue, it acknowledged the crisis's financial impact. More than ninety-five thousand claims from residential, business, and government entities represented a 90 percent response, huge compared to the normal 35 percent for class action settlements.[10]

Did the class action suit help reduce the water system's vulnerability to future spills? There was no immediate benefit because, similar to the PSC settlement, American Water Works Company was again allowed to seal expert witness testimony, prevent cross-examination, deny any liability, and blame the crisis on Freedom Industries.[11] A longer-term benefit, however, is promising though indirect: the settlement showed that, even in the heart of Chemical Valley, the water company was expected to find a way to provide an uninterrupted flow of safe water to its customers, regardless of the condition of its primary source water. Significantly, it had to pay customers for their losses when it fell short. As water activists discovered early in the crisis, the design of resilient water systems is not precedent setting.[12] An excellent example within our own state borders is the Morgantown Utility Board, where total organic carbon analyzers monitor primary and secondary raw water sources and their treatment plant routinely shifts between them to optimize its performance.[13]

From a longer-term perspective, the crisis's location and timing were fortuitous: its chaos coincided with the beginning of the regular state legislative session in Charleston. Legislators, lobbyists, advocates—everyone—were in town, and the regional crisis made immediate statewide and nationwide impressions. The special timing of the crisis, together with massive input from key legislators, water advocacy groups, and other water experts, produced Senate Bill 373, which was passed unanimously by the legislature and signed by the governor in record time.[14] Key provisions of this flagship legislation (1) initiated vigorous regulation of aboveground chemical storage tanks, (2) required public water systems to prepare source water protection (SWP) plans, and (3) established a Public Water System Supply Study (PWSSS) Commission charged with annually reviewing the new SWP Program and recommending improvements. This legislation can help achieve safe water systems essentially invulnerable to chemical spills. Even more beneficial, given the extraordinary set of circumstances accompanying the crisis, a stroke of the governor's pen extended its positive impact to all of the state's 125 public water systems.

Point 1 of SB 373 established aboveground chemical storage tank inspections implemented by the State's Department of Environmental Protection. Freedom Industries' leaky MCHM tank and the secondary containment around it were in deplorable condition. Its tanks had not been inspected for at least ten years, which underlined the critical role of tank inspections as part of safe water systems, those worthy of trust.[15] Perhaps it is appropriate that Randy Huffman, secretary of West Virginia's DEP during the water crisis, increasingly valued the peoples' trust in government and its relationship to transparency. Not surprisingly, he described the water crisis as "a turning point in some of his thinking about the DEP and about how environmental policy gets made in West Virginia." During this time he also reached the conclusion "the whole notion of transparency, it's an over-utilized, under-practiced thing. . . . The whole Freedom of Information Act is about the public having a right to this information. Give it to them. Give them what they ask for. Give them what they want. Give them what you think they want." He went on to say, "being open with the media and transparent to the public is all about integrity, and once government agencies and officials start to even appear to be hiding something, they lose any public trust they might have had."[16]

Point 2 of SB 373 required the state's 125 public water systems to prepare SWP plans. These compel water systems to understand potential contamination sources in their watersheds, to create emergency plans that anticipate chemical spills and to review their SWP plans—including public comment—every three years. They also require the water systems to assess water analyzers and alternate raw water supplies. These two components, a secondary water source and continuous water analyzers, are central to the design of safe water systems that continue to produce clean, potable water without interruption, even when their primary source of raw water has been contaminated.[17]

Point 3 established the PWSSS Commission charged with annually reviewing implementation of the SWP Program and recommending improvements. The commission comprised government, water treatment, engineering, health, and business representatives as well as a member representing public interest groups and a community member. Despite its lack of resources and authority, the commission was important because of its diverse background and because it provided, nominally at least, an overarching statewide review and feedback mechanism for the SWP Program. At its core was the mandate to consistently improve West Virginia's public water system by evaluating its performance and suggesting changes in practices and, if necessary, in West Virginia State Code.

Unfortunately, it was required to sunset June 2019. Missing here was an appreciation that the commission's review and feedback function was integral, not ancillary, to the SWP Program. Its positive impact was obvious when it contracted with the Horsley Witten Group, an unaffiliated party, to conduct the first evaluation of the SWP Program at a stakeholders meeting in 2018. The meeting produced detailed assessments and candid discussion on positive and negative features of the SWP Program. The commission's positive impact is seen in estimates from West Virginia's public water systems of the cost of alternative water sources, found to total at least $380 million, a large number but one that allows the rational planning of next steps in this dimension of the SWP Program. In its 146-page final report, the commission has documented these initiatives, the current status of West Virginia's SWP Program as found in the Horsley Witten Group's evaluation, and more than fifteen specific ways to improve it, most of which do not require additional investment.[18]

Strong public advocacy and legislative leadership must work together to revive the PWSSS Commission and act on its recommendations. This is crucial for the SWP Program to achieve its potential. Hopefully, another water crisis is not needed as a reminder that, although resilient water systems are much better understood and substantial progress has been made, West Virginia's public water system is not yet the one envisioned in SB 373.

Powerful voices to reduce the role of government and influential ones to consistently grow profits for the corporate component of West Virginia's public water system compete with the persistent notion that clean water should be a human right independent of one's income. But because clean water comes at a cost, footprints of this debate will persist in state policy. Since the passage of SB 373 in 2014, for example, at least two well-coordinated efforts led by West Virginia's extraction industries have reduced its scope, SB 423 in 2015 and HB 2811 in 2017.[19]

Overall, however, the 2014 water crisis was persuasive, and by 2019, SB 373 together with the stubborn support of citizen, political, and nonprofit advocates for clean water, has achieved much: SWP plans are in place for all of West Virginia's 125 public water districts, and the associated aboveground chemical storage tank program has registered 40,000 tanks, 4,000 of which are within five stream hours of a water treatment plant and inspected at the rate of 1,000 annually.[20]

At a rally in the capitol rotunda marking the third anniversary of the water crisis, Barbara Fleischauer, house delegate from West Virginia's Fifty-First District, remembered the 2014 crisis and the legislation it inspired. "One of the few upsides [to the crisis]," she said, was that "we got the regulations statewide that will help us in the future, that will make it less likely that one of these gigantic accidents can happen and there won't be anything in place. . . . So in my county [Monongalia], we have our water intake on the Monongahela River. We have assessments of the contaminants and we are working on a backup water source, all because of the horrible things that happened here [in Charleston and surrounding counties]. All over the state, people are working on these SWP plans. The legislation is not perfect, there's too much secrecy and too few contamination sources being included. But we're moving forward. Anybody who's been around here for a long time knows you move forward and back and keep trying to push it forward again. We moved *really* forward [with SB 373], and then we moved back [with SB 423], and we need to keep pushing it forward and making sure that these source water protection plans actually are implemented at the local level. . . . So, as scary as it was a few years ago, the people in this room have a lot of reason to be proud. . . . You participated in democracy. . . . You changed the law for the better. . . . I am very proud to have played a small part in that. . . . Thank you, all of you, . . . and don't give up!"[21]

Can we trust the water system in the nine-county region impacted by the 2014 water crisis? What about other water systems in the state—are they safe now? By definition, if something is "safe," it is "trustworthy, reliable."[22] It follows that, if our water system is safe, we can trust it day in and out, from one season to the next. Not surprisingly, the water crisis demonstrated in bold relief the definition of a safe water system: one that minimizes the occurrence of chemical spills but when they do occur can withstand a chemical challenge even more toxic and less odorous than that of January 9, 2014, and continue to produce safe water without interruption.

David Chairez, an interviewee met in chapter 8, spoke to this issue by gauging the public water system's ongoing performance: "How can I really trust that everything is OK, that I'm going to be OK, that my kids are going to be OK if we continue to use and consume the water. . . . For me, personally, it's a matter of time. I think it's a matter of consistency."[23]

This summary of quasi-judicial investigations, court proceedings, and legislation suggests that our own public water system, though not yet implementing continuous water monitoring coordinated with a second raw water source, is less vulnerable to chemical spills simply because the Freedom tank farm is gone. Also, the PSC General Investigation and the federal class action suit have further nudged WVAW, by direct and indirect means, toward fulfilling its role as a safe water system. In addition, SB 373 mandates a combination of actions and studies, which, combined with periodic, third-party evaluations of the SWP plans like that conducted by the Horsley Witten Group, should further drive WVAW and relevant state agencies to fashion a rigorously safe water system here in Chemical Valley. An extra bonus is that these same mandates now exist statewide, too, and stand to benefit an even larger population.

Measurable progress has been made, but much work remains: a critical feedback loop in the SWP Program, the PWSSS Commission, has been terminated, and industrial voices continue to argue that SWP guidelines are excessive and the legislative support required to implement needed improvements in the SWP Program seems to have dissolved. But recall that a strong citizen component was present in all the crisis components discussed in this chapter. A resurgence of citizen and nonprofit input and inquiry—in a word, activism—must combine with a new wave of legislative commitment to clean water as a basic human right. Together, these forces are strong enough to reinstate the review and feedback function of the PWSSS Commission to the SWP Program and to insist that its well-defined initiatives be enacted. In such an environment, our public water systems will continue improving until safe water systems are the only kind to be found in Chemical Valley and throughout the state. Barbara Fleischauer's admonition is still timely:

"DON'T GIVE UP!"

## Notes

1. Reginia S. Lipscomb, personal communication with Jim Hatfield, July 14, 2017; James Thomas, email communication with Trish Hatfield, September 25, 2017.
2. U.S. Chemical Safety and Hazard Investigation Board, Investigation Report: Chemical Spill Contaminates Public Water Supply in Charleston, West Virginia, Report No. 2014-01-I-WV (Washington, D.C.: U.S. Chemical Safety and Hazard Investigation Board, 2016), 19–22, 118.
3. Kate Mishkin, "Clendenin Businessman Buys Cleaned-Up Site of 2014 Chemical Spill," *Charleston Gazette*, December 14, 2018, A1.
4. See, e.g., Rusty Marks, "Containment Trench Overflows at Freedom Site on Elk River," *Charleston Gazette-Mail*, June 13, 2014, A2; Ken Ward Jr., "Runoff Trench Overflows—Twice, Incidents Thursday, Friday Just the Latest in a String of Mishaps at MCHM Leak Site," *Charleston Gazette-Mail*, June 14, 2014, A1; and Ken Ward Jr., "Freedom Proposes Runoff Control Steps, Changes due to DEP Orders, Water Company Criticism," *Charleston Gazette-Mail*, June 15, 2014, A1.
5. David Gutman, "Water Crisis Freedom Execs Linked to New Chem Company, Lexycon Shares Names, Numbers with Firm behind MCHM Leak," *Charleston Gazette-Mail*, May 1, 2014, A1.
6. Ken Ward Jr., "W.Va. American Water to Be investigated," *Charleston Gazette-Mail*, May 22, 2014, A1.
7. Kate White, "Lawsuits over Spill Moving Forward, Dozens of Suits Consolidated," *Charleston Gazette-Mail*, May 22, 2014, A5.
8. See, e.g., Ken Ward Jr., "WVAWC's Response to Spill Investigation Concerns State PSC," *Charleston Gazette-Mail*, August 6, 2014, C1; and Ken Ward Jr., "PSC Deal Might Lead to 2nd Water Intake," *Charleston Gazette-Mail*, January 26, 2017, A1.
9. Ken Ward Jr., "Judge Gives Water Crisis Settlement Preliminary Approval," *Charleston Gazette-Mail*, September 21, 2017, updated October 15, 2017, https://www.wvgazettemail.com/business/judge-gives-water-crisis-settlement-preliminary-approval/article_03e81eef-1569-5d19-bccb-52ad171814be.html.
10. See WSAZ news releases from April 27, 2017, to January 3, 2019, documenting progress in the federal class action lawsuit covering the spill of crude MCHM chemical into the Elk River in Charleston, Dan Klein, "Update: Nearly $3M in More Water Crisis Settlement Checks to Be Sent," WSAZ News Channel 3, https://www.wsaz.com/content/news/151-million-proposed-settlement-filed-in-2014-water-crisis-420683453.html.
11. Kyla Asbury, "Women Pregnant during Water Contamination Crisis to Get Checks Soon," *West Virginia Record*, January 1, 2019, https://wvrecord.com/stories/511713110-women-pregnant-during-water-contamination-crisis-to-get-checks-soon.
12. Philip Price and Jim Hatfield, op-ed, "There's a Better Way to Monitor Water for Pollutants," *Charleston Gazette*, April 29, 2015, A5.

13. Greg A. Shellito, Morgantown Utility Board, West Virginia, personal meetings and communications with Jim Hatfield, 2015–19.

14. See, e.g., Eric Eyre, "Tomblin Signs Water Tank Bill," *Charleston Gazette-Mail*, April 2, 2014, C1; State Delegate Barbara Fleischauer, op-ed, "Water Bill Was a Good Beginning," *Charleston Gazette-Mail*, April 10, 2014, A4; Ken Ward Jr., "One Year Later: Where Things Stand since the Leak," *Charleston Gazette-Mail*, January 9, 2015, A2; Randy Huffman, "On Freedom Spill Anniversary, Looking Back but Moving Forward," *Charleston Gazette-Mail*, January 9, 2015, A4.

15. CSB, Investigation Report, 36. See also Ken Ward Jr., "Feds: Freedom Knew of Problems for Years," *Charleston Gazette-Mail*, January 1, 2015, A7.

16. Ken Ward Jr., "'It's Not about You, It's about Your People,' Outgoing DEP Chief Reflects on Eight Years with Agency," *Charleston Gazette-Mail*, January 15, 2017, A1.

17. Jim Hatfield, op-ed, "Five Elements of a Safe Water System," *Sunday Gazette-Mail*, March 1, 2015, C1.

18. Final report of the Public Water System Supply Study Commission to the Joint Committee on Government and Finance of the West Virginia Legislature, December 17, 2018, https://wvrivers.org/wp-content/uploads/2019/01/2018 -PWSSSC-Final-Report.pdf. An evaluation of West Virginia's Source Water Protection Planning Program, July 2018, prepared by the Horsley Witten Group, Inc. is part of the final commission report and begins on p. 36.

19. SB 423: Ken Ward Jr., "House Passes Tank Bill," *Charleston Gazette*, March 14, 2015, A1; and Ken Ward Jr., "Chemical Tank Bill Signed," *Charleston Gazette*, March 28, 2015, A1. HB 2811: Editorial, "Limited Tank Exemptions Proper for Oil and Gas," *Charleston Gazette*, March 14, 2017, A5; Ken Ward Jr., "House Tanks Chem Bill; Lawmakers Balk at Larger Rollback of Storage Law," *Charleston Gazette*, March 14, 2017, A1; Ken Ward Jr., "House Passes Tank Bill Regulation Rollback Scaled-Down Version," *Charleston Gazette*, March 18, 2017, A1; Ken Ward Jr., "Tank Safety Roll Back Passes, Water Pollution and Drilling Bills to Come," *Charleston Gazette*, March 26, 2017, A1.

20. Final report, Appendix E, 33.

21. West Virginia Delegate Barbara Fleischauer's remarks at public gathering marking third anniversary of water crisis, state capitol rotunda, January 20, 2017, video on Facebook page of West Virginia Rivers Coalition.

22. *Merriam-Webster Dictionary*, merriam-webster.com.

23. David Chairez, interview with Jim Hatfield, August 5, 2014; with revisions, personal communication with Jim Hatfield, July 25, 2017.

# Epilogue

Luke Eric Lassiter

In late January 2018, a group of those involved in this project met in Charleston, West Virginia, to discuss a draft of this book. Gathered were several of the book's authors—namely Trish and Jim Hatfield, Cat Pleska, Beth Campbell, Brian Hoey, and me—as well as several of the oral history consultants involved in the work from its earliest stages, including Carla McClure, Linda Koval, Paul Sheridan, and Paul Epstein (who joined us via conference call), and Renate Pore, who, when we started this project, served as the board of directors president of the West Virginia Center on Budget and Policy and who first encouraged us to collect water crisis oral histories as a project with the center.

Around the same time four years before, many of us in the room had just begun to use our water again after completing the flushing process, but even then the crisis was far from over. As this book demonstrates, for the next several years, trust in our local water system became a critical concern. As we prepared to talk about the draft manuscript that evening, I couldn't help but think that just two years prior, in 2016, Beth and I had started using the water in our home again for cooking and drinking, though even at that time we did so only periodically. Indeed, then as now, with history in mind, I wasn't so sure if I could ever completely trust the process on the other end of our home's water taps. Of course, from our work on this book, I knew that others here and throughout our community shared similar sentiments.

All of those present for the discussion had read much or all of the draft, some more than a few times in its various iterations. Our purpose was twofold. First and foremost, we sought to identify and discuss any glaring mistakes found in the manuscript and to make sure we had things "right," as it were, or as close to right as possible. Several participants, like Carla McClure and Paul Epstein, provided detailed, written comments during and after the meeting, which were, of course, extremely helpful. And some, like Linda Koval, provided further clarifications on quotations. This work of checking and clarifying quotations had been ongoing since the past summer, when members of the original oral history research team (which included Marla Griffith again, along with Cat, Jim, and Trish) had begun the process of sharing parts of the developing manuscript with everyone interviewed for the project.

As I outline in the chapter 3, these kinds of so-called participant checks are an important part of doing a collaborative ethnography couched within an ethics of representation that takes seriously the experience and expertise of so-called informants as—alternatively—consultants, who are cited and engaged as named contributors. But it is only a first step. An equally important part of this process is to engage with alternative readings and interpretations, which involved the second reason why we gathered to discuss the manuscript as a larger group. Other readings and interpretations can and often do shift the design and development of an ethnographic manuscript as it begins to take shape. This was certainly the case with this work. At the same time the research team was engaged in the process of participant checking, Beth, Brian, and I were reworking the manuscript into a more coherent whole. And as comments came back to us from consultants via the oral history team, the manuscript began to solidify and take a form somewhat similar to its current form.

So it was that on this evening in late January, with a more or less final manuscript in hand, that our small group gathered to engage in a collective reading and discussion as we prepared the final manuscript for its submission to West Virginia University Press. I began our meeting by reminding everyone of our purpose, that as participants in an ethnography, we were engaged with reading a particular kind of literary genre that describes the down-on-the-ground experience of an event or place. And in this sense, our book necessarily and consciously—as all ethnography—would be partial and incomplete, a book not meant to be, for instance, investigative journalism or policy research or a kind of case study found in other genres of social science research. It is, indeed, meant to fill in voices often not heard (deeply, in any case) in the important work already carried out by investigative journalists or policy and other social scientist researchers who, for example, have focused more on the "big players" such as Freedom Industries, West Virginia American Water, or the State of West Virginia.

I also reminded everyone that because this was a more or less collaborative ethnography written by both academics and community authors, the book addressed audiences both academic and local, as well as, more generally, a general reading public inside and outside of West Virginia who might be interested in the issue of safe water. Thus our more overall purpose here was to assess the extent to which we were able to achieve that which we had set out to do: again, to elaborate the experiences and perspectives of those who encountered—and then dealt with—this particular environmental crisis firsthand.

After introductions and some conversation about the layered approach to storying the water crisis that ran alongside the research and writing process,

Renate spoke directly about the text's content. She began by describing how the larger players of the crisis help to contextualize the work. "The book isn't just about the water crisis," she began; "it's about West Virginians, how they feel . . . , and about how disappointed people are—not just with what happened with the water, but with being let down by the governing class here. I think you did a good job of pulling coal into this: the politicians, you know, said, 'this has nothing to do with coal'—yes, *well*, it *does* have something to do with coal. It has something to do with the disrespect that Freedom Industries, West Virginia American Water, or the coal industry have for people who live here, [industries whose leaders] who feel like they can get away with anything.

"But one thing that I didn't see, but perhaps we can say that Jim takes this up some in his chapters . . . , was that you didn't take on West Virginia American Water. . . . I thought you let them off the hook pretty easy. It's not just the way they reacted to the water crisis, but the decisions they made before the water crisis ever happened: they didn't do due diligence in the way they drew their water down. . . . No one should have just *one* source of water; they should have been able to block [the MCHM leak] off and go to another source of water." Renate spoke about a friend who managed another local water facility and his experiences with trying to convince officials to include another water source for Charleston's facility. She then returned to the topic at hand: "I just think you could have been a lot harder on them."

As Renate spoke, I thought to myself about how during the early days of the spill, WVAW seemed more like a victim than a culprit. But as the larger water crisis emerged and as other associated issues became apparent (such as not having an alternative water source), many began to see WVAW in the same light as other industries—such as coal—whose negligence seemed to help usher in the crisis in the first place. Indeed, a recent class action settlement in which WVAW agreed to pay damages to its customers was fresh in all of our minds. Paul Sheridan, for example, had followed the case closely as the attorney for Advocates for a Safe Water System. I thus couldn't help but think at that moment that this evolving view of WVAW had certainly affected how we first approached the oral histories, then how we wrote the book as it first materialized and as it evolved into its current form. I was very curious about what others gathered here might say about Renate's last point. "How do the rest of you react to that?" I asked.

Jim, who, as Renate suggested, covers some of this issue in his own chapters, was the first to respond: "From my point of view, the focus of the book was more on the people who were interviewed and how they felt about the crisis. It's certainly true that a lot more could be said about the role of West

Virginia American Water." Jim turned to an even larger issue for him: "But what brought me into this whole thing was—you use the word *disrespect*—a sense of arrogance that I heard, combined with people talking about technology who didn't seem to *know* about technology. Those two put together, plus all the pain it put people through: that got me very excited and upset." Jim turned to Paul Sheridan, who sat next to him, and asked, "Do you have something to say about the role of West Virginia American?"

"Well," started Paul, "my sense was the book really was more about the experiences of the people you talked to. One of things that I really liked was that you focused on what people had to say about *trust*—that's something that went through people's stories here. Which is a *huge* factor . . . parallel to social capital, in the sense that it's an asset, albeit an abstract asset: if people trust things, then things work."

"That's the 'speed of trust,'" interjected Carla, referring to Stephen M. R. Covey's book by the same title. "Covey talks about how if you trust one another, then you can skip all the preliminaries and get things done."

Our conversation was becoming more lively now as many in the room began nodding their heads and interjecting sentiments of approval. "Yeah," said Paul, nodding his head in agreement and speaking in a more animated tone. "I don't know the specific concept, but I think that's part of what social capital is: linkages between people that work, and that people can rely on. . . . In the early stages of the water crisis, the lack of trust was a sense of betrayal. There wasn't a lot of information, things weren't transparent, and there was a lack of understanding, and lots of technical issues. But now I think we know more that supports people's suspicions, and confirms a lot of the distrust that people may now have."

"I have a new perspective on all of this," said Linda, expressing a feeling that several of us seemed to share. To be sure, our views on the causes and scope and implications of the crisis had evolved over time.

Our conversation turned to talking briefly to the private versus public ownership of local water systems. I then returned to Renate's original point, noting that while I thought that many others, like the *Charleston Gazette*, had covered the WVAW issue well, I still very much appreciated her raising the issue because it helped to clarify the choices we had made in deciding to write the book the way we did, emphasizing people's experience perhaps at the expense of larger players like the local water company. Indeed, the agreements and differences that helped to give life to the oral histories and then the book had changed significantly over time and were still changing. Our conversation was clearly illustrating that.

"But there are so many other things, too, that I thought I might encounter in this book that weren't there," said Carla after a short while. "You just talk in general terms about the spin that the government agencies were putting on things. But I would have liked to have seen a couple of quotes [along these lines] because if you didn't live through it and vividly remember those people saying those things like I still do, then you might not realize exactly what those things said were implying." Carla had made the same point in a written list of suggested changes, and I made some notes about incorporating some quotes back into the manuscript to illustrate her points. The task seemed pretty straightforward, I thought, and very doable at this point in the manuscript's development.

"Saying things like what?" asked Jim.

"Like when [bureaucrats] were saying—and the governor picked up on this—'Oh, all these people going to the emergency room with symptoms: it's just something like a flu or anxiety.' I remember feeling sick at the time, and I was watching this on TV, and I got *so* mad. I got out of my chair, stomped around the room, and was screaming at my husband: 'These people! Are they really going to expect us to sit here and believe this stuff!'"

As Carla spoke, I remembered feeling the same way when, perhaps following in the vein of various "official" dismissals like those Carla articulated, a friend disregarded Beth's stomach pains the day after the spill as some sort of anxiety. That was possible, of course, but so, too, was it possible that she'd had a physical reaction to the MCHM, which our friend—along with governmental and company officials—was slow to acknowledge. For several minutes, we talked about Carla's and others' sicknesses, as well as other experiences associated with the water crisis, being repeatedly disavowed, and how we might go back and emphasize this more in various parts of the draft manuscript.

At this point Paul Epstein chimed in from across his conference call connection, pointing out that "the book is mostly an ethnographic study . . . , an effort to hear the people's story and to hear it from different perspectives. So it doesn't give you *one* perspective; it gives you many perspectives, and it's up to the reader to put them together. Maybe some other books might treat some of these other issues . . . but when you get into people's perspectives, you don't necessarily get *all* the facts. . . . In this case, anybody who was there and lived through it, you're going to see in the book in terms of what you experienced, and *that* you remember as fact. I can't know what it'll be like to read this book from the perspective of someone who wasn't *here*. Brian, though, tries to give us a much bigger picture of disasters in other parts of the country, and other parts of the world, and then tie them into how our water crisis relates to those and how it affects people, too. I think there are a lot of interesting things in this book. It's an *unusual* book in many ways."

On Paul's last point, conversation turned to the unusual style of the book and its somewhat nonconventional arc, with multiple and variously situated authors, stories, and voices. Carla, for example, suggested that we do more to summarize each chapter, perhaps adding a brief abstract or summary at the beginning of each chapter. (We added these in the first draft sent to WVU Press, but, per peer reviews, moved them, in part, to Brian Hoey's interlude in the final manuscript). Others thought we might develop more explicit cues (like adding more references to relevant chapters throughout, which we also did) to help the reader navigate the overall work. Or how we might address points of dramatic shifts in the overall tone of the text (a suggestion that prompted revisions to the introduction).

This part of our discussion led us to talking about Brian's chapter (now split into a chapter and an interlude), which is perhaps the most academically styled discussion in its comparative breadth and depth. Indeed, Brian invites the reader into a complex and ever-changing landscape of disaster studies, with which even seasoned scholars can often struggle to keep current. From the beginning of our project, Brian worked to provide our team with a framework within which we could place—and then better understand—our developing locally based work. But he was also sensitive to how his chapter might be seen, by some readers, as a kind of "final say" about how to analyze the experiences of those featured in the book. Many ethnographers today, like those of us involved in this project, are cognizant of how local experience in these types of community-based research can be effectively "colonized"—or "mined," a term used often here to describe both literal and figurative exploitation—and then appropriated and forged into something they are not to serve outside research agendas. Be that as it may, though, the goal in collaborative forms of ethnography is not necessarily to replace academic perspectives with community-based ones but to put these differing perspectives into dialogue with one another.

Brian spoke some about the issue directly: "I didn't want to *set* some kind of interpretive frame for everything, to say, 'this is how those people experienced this' or 'this is how you should interpret those experiences' or 'this is an instance of this or that experience.' But rather, again, I wanted to provide a broad context, so that the reader can say, 'Oh, I can *see* this; this is something I've already seen.' So I'm very *careful* in how I've done it, and if I should do it differently, I would be more than happy to think about this more.

"For my own purposes," Brian continued after a short pause, "I went through every bit of every transcript and found connections, and made connections, in my own mind. I have an incredibly detailed list of every single instance of those things. And then I ignored all of it!" Brian smiled and laughed slightly, his tone now inviting a mix of intellectual play and careful reflection. "Then I went to the literature and said, 'Okay, having steeped myself in all of that, what speaks to me in the literature? And will allow *me*—hopefully by some process of having made those connections—[to] also make connections for other people, or allow for them to make connections that can be made. And if they don't make them, well, they don't make them.'"

"Yes, I felt 'popcorn' going off in my head the whole time I was reading your chapter," said Carla. Several folks smiled and nodded in agreement.

"That's good," said Brian, continuing with the metaphor in a humorous tone. "I'm *looking* for popcorn—homemade, 'make your own popcorn.'"

Our conversation grew more involved now, with several folks contributing their own thoughts about and suggestions for Brian's chapter and about how we might further address this tension between letting people speak for themselves and, at same time, having someone else analyze their experience from a more or less analytical or outside position. Both kinds of expertise, after all, were critical to our project. We talked more about this tension for some time, coming back, with this in mind, to reframe our exchange about issues of trust, the role of WVAW, the book's audience, and how the book fit into an ethnographic genre, even with its many different styles and approaches ranging from oral history to memoir to analysis. But this talk inevitably turned us toward a more immediate issue encapsulated throughout this book: that between inside and outside, or, more specifically, between how outsiders often perceive West Virginians and how insiders understand and experience West Virginia.

Oftentimes this tension surfaces as a struggle between, on the one hand, the stereotypes of poor, backward Appalachian folk that outsiders often use to describe life in West Virginia and, on the other hand, reactions to these stereotypes by West Virginians themselves who endeavor to present more positive images of life in Appalachia. But there is a more complicated and nuanced side to this discourse of image. Emily Mayes's chapter ("In and Out of Appalachia") came to the fore on this point. "I really enjoyed reading about her experience," said Trish, after introducing the issue. "And I certainly would not deny her the voice to do that. She acknowledged systemic challenges that I don't think anybody else covered to the extent that she did. But the last three pages, to me, felt like she got into victim-blaming with the phrase 'tradition of complacency.'

I work with nonprofits . . . and they are *not* complacent by any means. So my exposure to that kind of set off alarms."

Trish felt strongly that we, as a group, needed to acknowledge how to engage other ways to think about this issue, such as the effect of sustained poverty and how "if you live in generations with this, and or if you have these experiences of not trusting things, then that would dull the notion that 'well, yeah, I need to go do something.'" Indeed, Trish insisted, "traditions of complacency" didn't tell the whole story. Trish's point propelled us into even more involved conversation as we turned to talking about the context in which Emily wrote her chapter: in a graduate seminar that included internal critiques of the state (such as that by Eric Waggoner, featured in chapter 5) and then in the context of Emily's own experience growing up in West Virginia, wanting to stay here but not being able to, given the lack of opportunities for many young professionals. And then when the water crisis hit, for many young people like Emily, that was the last straw: Why stay?

"So it's a perspective that's part of all of this," I said, wanting to be clear about why I encouraged Emily to write the chapter in the way she did.

"Matthew Chesebrough, who I interviewed," said Cat, jumping into the fray, "also hit upon that." Referring to Matthew's anger about how young people's concerns are so often dismissed here, Cat continued, summoning Matthew's voiced sentiments: "'I'm thirty. You don't care about me. You don't care about how I'm going to have a future here.' He was *angry*, too. He didn't express it in the same way that Emily did, but I noted the similarities between them. They are near the same age . . . so I wasn't surprised when that came out in her chapter, that she was that angry. But she does represent a demographic . . . because some younger people feel disenfranchised here."

With this in mind, our conversation eventually turned to citing the people we knew—such as Emily and Matthew—who, prompted by the water crisis and how it was handled, left the area, frustrated and angry, perhaps for good, and, of course, those who stayed, for a variety of reasons (some of which we cover in previous chapters). Trish, in particular, reminded us again to remember the many people, old and young alike, who work daily to make life better here, even given the state's very real and serious challenges. "For some of us, *because* it's so challenging here, we find a greater life here."

"It seems to me," said Paul Sheridan, "that one key theme in the book is this question of what I would call hopefulness versus despair. It's an even bigger issue here. Trust is a piece of it, but it's bigger than that. It's a big theme for this place right now, what we're talking about with Emily's chapter and with her feelings. I, too, feel sometimes like I live right at the crux of that. . . . I feel like I go back and forth. That teeter-totter was accentuated by the water crisis. I had moments of feeling in the midst of the crisis, thinking, okay, this could be our finest moment. . . . In the book, people in the interviews, people were saying that 'people are going to rise up and do the right thing. This could be the moment for us to take control of the water system.' Ten years from now, looking back, Charleston could be the place that totally overcame that, and became the example for the rest of the country. But I don't know if we're on that track at the moment."

We wondered together about Paul's last point, considering the many positive things that had come out of the crisis, as well as its many negative consequences. For most of us gathered, and for many of those we had come to know better through this project, things had clearly improved, and trust was perhaps growing again. For others, though, not so much. As our conversation continued to unfold along these lines—and as Trish shared many inspiring and hopeful stories of local people and nonprofits working hard to change things, to make life better, and as others shared knowledge of people who had left the area over this issue, even our own Jay Thomas (author of chapter 5), who had by this time sold Blues BBQ—we seemed to find common ground that this issue remained an open question and that both sides had value in trying to understand the diverse long-term effects of the water crisis.

"What she is saying *is* true," said Renate, returning to Emily's chapter. "But so is the opposite; it's *just* as true."

Paul Sheridan came back into the fray, referencing the "hopeful, optimistic" side of this equation: "Some days I feel like I can go with that, and then other days, I feel like . . ." Paul's voice faded as he shrugged his shoulders, seeming to imply the rest of the sentence: "giving up."

We talked briefly about how so many "long-haul" activists—like many in the room—remained positive and the take by some here that activism was growing in the state especially among young people, for the first time in decades—in ways that it had in the 1960s and 1970s. I referenced another one of our program collaborative research projects, the West Virginia Activist Archive, where this certainly seemed to be the case. Trish, Renate, and others provided additional and powerful examples. Several of the most important nonprofits in the area, like the West Virginia Center on Budget and Policy, were staffed by incredibly capable, articulate, and inspiring young people.

Yet part of the frustration of some young people, like Emily and Matthew, said Beth, includes "that if you want to come here and be an activist, you can do important work. But what if you don't want to be an activist? What if you just want to have a life?"

"And be with your family," I said, thinking about many of my own students in their twenties and thirties, raised in West Virginia, who had expressed that sentiment to me more than a few times.

"And you just want a job," said Renate.

"Yes," said Beth. "And you just want a job. What if that's what you want, and not the activist's struggle?"

"I think that's a fair statement," said Cat, nodding her head.

Silence briefly fell across the room for the first time that evening. Cat's comment returned us to acknowledging how important it was to feature voices from those like Emily and Matthew, as well as the important voices of activists who were continuing to carry the flame. Perhaps, after all, it wasn't an issue that we could easily resolve in a conversation that now was approaching its second hour.

Realizing this, I thought it important to start winding down and began to say so. But Paul Sheridan wanted to insert a final thought: "It seems to me, [on] this final question about the water crisis, and the question about those who don't want to be activists. . . . I think the water crisis raises, for me, a deep question about whether that is a viable position. I get that everybody, at some point, doesn't want to have to always be engaged, fully, in trying to change things and hold people accountable. But I think part of the lesson of the water crisis was that, on the whole, we learned that we have to do a whole lot more of that than we do now."

"You mean being engaged?" Jim asked.

"Yes," said Paul, nodding.

*That*, I thought, nicely indexed an important point about the draft manuscript and our conversation that had materialized that evening. But it also, in part, summed up our evolving process. In many ways, the water crisis had turned many of us into more engaged citizens in one way or another, whether we had wanted that or not. And this included the research and writing that gave life to this book. I couldn't help but think at that moment how the energies and passions of Jim and Trish had compelled me to get more involved in the crisis and its implications nearly four years earlier. Before that lunch with Trish and Jim at Blues BBQ, I just really didn't want to get involved. I was angry about the MCHM contamination, to be sure, but I just really wanted to do my job at the university and move on with my life. At that time, I admit that in the weeks before the project began, I dreaded doing another research project that would take years to complete (as this one certainly has) and that may or may not have any impact. Reflecting now on Paul's words, I realized how dramatically my thinking had changed about this project since then, and how far I—and we—had come. Indeed, the water crisis and the stream of action and even activism into which it pulled us, had led to this conversation this evening, to discussing tensions and issues that deeply mattered to us. Perhaps, in some small way, it may have also led us toward working together to make a difference—as citizens, many of whom, of course, just want to get on with their lives but who also need and want clean water. This, above all, is why we wrote this book.

# Afterword

Angie Rosser

The Elk. The longest river that begins and ends in West Virginia. It starts as a merging of springs in our mountain headwaters region, then meanders 177 miles through the heart of our state into Charleston, our capital city. I consider it my river. I live along its banks and am humbled each day that I get to gaze on it, ever flowing, ever changing in color and characteristics. For me it's a reminder of why I do the work I do with the West Virginia Rivers Coalition (WV Rivers).

WV Rivers was founded in 1989 by river enthusiasts distressed by noticeable degradation and increasing pollution threats to our rivers. Those founders of the first statewide water organization east of the Mississippi recognized that rivers needed a voice. Over time I've realized that we speak not only for rivers and streams but for the wildlife and the people that depend on them. During three decades together, as thousands of people speaking as one, WV Rivers has promoted protective water policies, confronted illegal pollution, and advocated for the conservation and restoration of our rivers for all of their uses.

I distinctly recall a conversation with a young man named Rob Goodwin when I first took the position as WV Rivers' executive director, well before the 2014 water crisis. Rob was an advocate for rivers and communities threatened by the environmental and health impacts of coal mining. He knew the state's river systems well. When I shared with him my affection for the Elk, he remarked that it was in pretty good shape. Especially when compared to other rivers in West Virginia, its water quality and biodiversity were reputable.

Rob worked in communities where drinking water contamination was the norm, so he was familiar with how drinking water was sourced and how things could go wrong. Speaking of the Elk, he noted how unusual it was that such a large swath of West Virginians—nine counties in all—depended on a single drinking water intake on the Elk. I remember thinking, "Well, good. It's reassuring to know that the river that supplies the largest customer base in the state is fairly healthy and stable." Indeed, I too was one of those people who took the Elk for granted.

The 2014 water crisis was a remarkable time to have the nation's eyes on the Elk—and on the consequences of failing to protect it. Across our state and beyond, many people wondered for the first time, "Where does my drinking water come from? What's in that river that ends up in my body?" It was a powerful realization of the connections between river protection, public health, and economic security. Clean, healthy rivers became recognized as a necessity.

The scale and intensity of people organizing in 2014 for stronger water protections was like nothing we'd seen before in West Virginia. It was a powerful opportunity to reform the way the state approached oversight and management of our rivers. Good legislation was passed, agencies were called to account, corporations were scrutinized—all because people were paying attention and getting politically active.

Still, I struggled with these outward measures of success. There were, and are, many West Virginians whose right to clean water has been violated for so long, and they remain invisible. There are places like Prenter, where residents had to pursue litigation against a coal company for contaminating their well water. There are similar stories now from the gas fields; "water buffalos" that have replaced fouled wells are a common sight. On any given day, people living in Gary don't know if the water coming from their taps will be colored rust, white, or gray. But those places aren't in the state capital; they aren't affecting three hundred thousand people at once; they aren't in the national spotlight. Yet in those pockets of West Virginia, the water crisis has been going on for decades.

So the truth is, the water crisis isn't over. Thousands of West Virginians live without access to clean water. Caity Coyne of the *Charleston Gazette-Mail* recently published a series of articles documenting numerous water problems across West Virginia. Nine public water systems have been on boil water advisories for five years or more. The boil water advisory for the community of O'Toole has not been lifted since 2002. Residents of Arthurdale receive regular notices that their water exceeds safe limits for cancer-causing chemicals. They are told to consult with their doctor to determine whether or not it's safe to drink the water.

It's maddening to know that many of these water problems affecting West Virginian's health could have been prevented. It's often overwhelming to process the costs of poor decisions made regarding our water resources across the state—just as it was disheartening to watch our government unravel progress made toward better protections for our water following the 2014 water crisis.

As highlighted in Jim Hatfield's writing, industry's success in lobbying for broad exemptions and loopholes has left the state's regulatory program for chemical storage tanks a mere shell of its original self. The West Virginia Public Water System Supply Study Commission's well-documented recommendations for essential public water system improvements across the state remain largely unfunded, unaddressed and ignored by lawmakers.

I'm often asked if the water is any safer than it was before January 9, 2014. It's undeniable that risks remain. In recent years the political push to deregulate has gained traction. The governor placed a moratorium on any new regulations. Agencies must report to the legislature any state regulations that are more stringent than a federal counterpart to ensure, for the benefit of polluting industries, that West Virginia provides only the minimum possible protections. All of this, I believe, is leading us to increased vulnerability for our water supplies.

And now? As the saying goes: out of sight, out of mind. As the Elk River chemical leak fades into history, so does our leaders' zeal for protecting water. I know, though, the victims of the 2014 water crisis haven't forgotten and will never forget. But when life carries on and day-to-day responsibilities and the immediate burdens that families face take priority, it becomes more and more difficult to expect people to stay engaged as they were in 2014. Those crowds aren't at the capitol. The outrage dissolves, and politicians get back to business as usual—until the next crisis.

West Virginia Rivers Coalition exists with the support of people who know the value of water and recognize that it's up to us to act to protect it. For our members, that was the big takeaway from the water crisis: we cannot be complacent; we cannot assume our water will be safe. We all have a responsibility to be informed and involved.

Right now, West Virginia's water quality standards, which limit the amount of toxins allowed in our water supplies, are under review. Some of our state's standards have not been updated in more than thirty years. The latest science and recommendations from the U.S. Environmental Protection Agency say that we should make updates to adequately protect human health. Yet, earlier this year, at the behest of the West Virginia Manufacturers Association, the West Virginia legislature said no. The chemical manufacturers dumping toxins in our water supplies say the updated human health protections will be too expensive for them to adopt. They say it shouldn't matter that our state allows more dangerous chemicals in our rivers than in other states because West Virginians weigh more, drink less water, and eat less fish.

The people who drink water supplied by West Virginia's rivers and streams deserve better.

Water is political. Rivers need advocates. Our role at WV Rivers is to bring citizen advocates together, to speak with one voice, the people's voice. We know that is what the Elk and all of our rivers need. We know that our power rests in people's commitment to pay attention and act. We are grateful for all those who are able to donate to the cause of keeping the people of our state informed. We don't want another water crisis to become that teachable moment.

Thank you for reading this book. On this walk through the experiences of West Virginians, you might have been struck by the thought that we are not unlike other places dealing with serious water issues. This is true. Increasingly, across the nation, across the world, access to clean, safe water is becoming the most urgent need of our time.

A slogan that emerged during the organizing around the water crisis was, "Water Unites Us." I understood this truth then, when I saw so many people from so many backgrounds come together for water justice. Today, I understand it more broadly. Water *connects* us—what happens to West Virginia's rivers affects everyone downstream. And water *unites* us.

We must count on one another to do as Wendell Berry reminds us: "Do unto those downstream as you would have those upstream do unto you."

# Acknowledgments

Soon after the Elk River Spill, and long before this book was even an idea, a group of us made an application to the Oral History Association's Emerging Crises Oral History Research Fund to document the experience of those on the receiving end of the spill and crisis. It was one of the very few grants that allowed for a quick turnaround to document a crisis or disaster. Because that award ultimately made this book possible, we want to recognize the critical importance of this program—and encourage other associations and organizations to develop similar funds.

Thus supported by the Oral History Association, our oral histories were carried out under the auspices of the West Virginia Center on Budget and Policy. We want to especially thank Dr. Renate Pore, then board of directors president, who initially encouraged us to do the oral histories as a center project, and the center's staff—particularly its executive director, Ted Boettner, and operations manager, Linda Frame—who managed the project's budget and other logistics from beginning to end.

The makeup and work of the original oral history team is documented herein, but we want to thank especially the many people who agreed to take time out of their busy schedules to share their stories, many of which we relay in this book. Because this is a collaborative ethnography, in most if not all cases, each interviewee reviewed her or his stories and checked quotations for accuracy; they often provided additional—and important—information and helped to deepen analysis and interpretation, which extended their role beyond being interviewees. This work required even more time on the part of our interviewees, and for that we are especially grateful. A few of these consultants also collaborated with the larger team to read and respond to an early manuscript draft (the epilogue chronicles a meeting of that group): thanks to Paul Epstein, Linda Koval, Carla McClure, Renate Pore, and Paul Sheridan for their exacting editorial comments and suggestions for improving the work.

Our work moved into Marshall University's graduate humanities program in the context of two consecutive graduate seminars, where much of the writing for this book took place. These were not traditional courses, however, with faculty and students sequestered behind university walls. Based at the university's graduate college campus in South Charleston, these seminars included several of the original oral history team members and others from our community interested in the crisis, in addition to faculty and students. A seminar classroom on this graduate campus can be a unique place, where open and regular partnership, reciprocal exchange, and collaboration between university and community is valued and encouraged. Without this kind of openness and support from Marshall University, this book would never have seen the light of day.

# Contributors

**Laura Harbert Allen** is an Appalachian media scholar and producer whose research interests include power, media, and knowledge production in Appalachia. She is also interested in how gender, race, and class play out in the media. Her production credits include the MacArthur Foundation, Inside Appalachia, and Making Contact.

**Elizabeth Campbell** taught at Marshall University from 2012 to 2018. She is chair of the department of curriculum and instruction at Appalachian State University. Her research explores the constitutive nature of collaborative research and writing and especially how it works—through shared agency, shared commitment, and shared humanity—to make and remake those who engage it. Her most recent collaboratively written books include *Re-imagining Contested Communities* and *Doing Ethnography Today*.

**Brian A. Hoey** is a professor of anthropology and associate dean of the honors college at Marshall University. His research encompasses themes of person-hood and place, economic change and identity, and environmental health. His most recent book is *Opting for Elsewhere* from Vanderbilt University Press.

**Jim Hatfield** has a PhD in chemical engineering from the University of Minnesota. He had a twenty-five-year career with Union Carbide as a research scientist. He became an advocate for safe water systems following the 2014 Charleston water crisis.

**Trish Hatfield** is program assistant for the Marshall University graduate humanities program and a board member of Step by Step, Inc. She recently retired her facilitating business so she could focus her attention on writing creative nonfiction and participating in collaborative ethnographic projects.

**Luke Eric Lassiter** is a professor of humanities and anthropology and director of the Marshall University graduate humanities program. He is the author of several books on anthropology and ethnography, including *Invitation to Anthropology*, *The Chicago Guide to Collaborative Ethnography*, and, with Elizabeth Campbell, *Doing Ethnography Today*.

**Emily Mayes** graduated from Marshall University in 2016 with an MA in humanities and a graduate certificate in Appalachian studies. She works as a high school English teacher in North Carolina.

**Joshua Mills** graduated from Marshall University in 2016 with an MA in humanities and a graduate certificate in Appalachian studies. He is currently working as an archaeologist and survey technician for an engineering firm in Maryland.

**Cat Pleska**'s memoir, *Riding on Comets* (WVU Press), was a finalist in the *Foreword Reviews* memoir category. She edited the 2019 anthology *Fearless: Women's Journeys to Self-Empowerment*. She is working on an essay collection titled *The I's Have It: Traveling Ireland and Iceland*.

**Angie Rosser** is the executive director of West Virginia Rivers Coalition, bringing a background of working in West Virginia on social justice issues in the nonprofit sector. She holds a BA in anthropology from the University of North Carolina at Chapel Hill and an MA in organizational communication from West Virginia University.

**Gabe Schwartzman** is pursuing a PhD in geography at the University of Minnesota and has produced several research projects about the Appalachian coal fields, including the interactive mapping project WVWaterHistory.com and oral histories of the Appalachian South Folklife Center and Blair, West Virginia, both housed at the University of Kentucky.

**Jay Thomas**, 2017 Marshall University MA in humanities graduate, is a restaurateur and lover of literature. He and his wife, Honor, are relocating to the eastern panhandle of West Virginia. Their daughter, Daisy, is an actress living in Brooklyn, New York, and their son, Jake, is a student at Shepherd University.

# Index